Shipping Container Homes For Beginners

The Ultimate Simple Guide to Building Your Eco-Friendly and Budget-Friendly Container Home

Alexander Harbinger

Table of contents

Foreword

In the quiet, unassuming corners of urban sprawl, a revolution simmers one that redefines the very foundations of housing. The concept of sustainable living, once the purview of the few, has cascaded into the consciousness of the many, driven by a growing awareness of environmental stewardship and economic necessity. This movement, characterized by a deep-seated desire to minimize one's ecological footprint, aligns itself not merely with conservation efforts but also with a profound reassessment of what it means to inhabit our planet thoughtfully and resourcefully.

Among the myriad expressions of sustainable living, the use of shipping containers as residential homes stands out as both a practical solution to housing crises and a bold statement of eco-friendly living. Initially born out of necessity and the availability of surplus containers, these homes have transcended their utilitarian origins to become symbols of architectural innovation and environmental mindfulness. As cities burgeon and the call for affordable housing grows louder, shipping container homes represent a nexus of opportunity affordable, yet undeniably green.

The architectural landscape is witnessing a transformation as shipping container homes challenge traditional paradigms of building and design. These structures, with their modular, Lego-like adaptability, offer a unique aesthetic that marries industrial chic with minimalist tendencies. The intrinsic strength of steel boxes provides a robust framework that architects and designers have leveraged to craft spaces that are not only functional but also visually striking. This blend of form and function appeals particularly to those who seek a simplified lifestyle that does not compromise on style or structural integrity.

The rise of container homes is more than a trend; it's a cultural shift towards minimalism and sustainability that resonates deeply with a broad spectrum of individuals from young, urban creatives to retired couples looking to downsize their lives.

Communities of container homes have sprouted up across the nation, each reflecting the shared values of its inhabitants: a commitment to living lightly on the earth while fostering a sense of community that traditional housing often lacks. These enclaves are not just places to live; they are microcosms of cooperation, shared values, and mutual support.

Opting for a shipping container home is a choice laden with environmental benefits. Reusing containers as living spaces is an exercise in creative recycling that significantly reduces the demand for new construction materials, thereby lessening the overall environmental impact associated with building. Furthermore, these homes often incorporate other sustainable practices such as rainwater harvesting, solar energy installations, and greywater systems, amplifying their green credentials. In essence, each container home serves as a beacon of sustainable living, demonstrating that personal choices can indeed have a positive impact on the environment.

Economically, container homes offer a compelling alternative to traditional housing. The relatively low cost of acquiring a shipping container coupled with reduced construction time and labor expenses renders these homes particularly attractive to those seeking affordable housing solutions. This economic accessibility does not only open up homeownership to a broader demographic but also provides a practical blueprint for addressing the urgent need for housing in overcrowded urban centers and beyond.

As we stand at the crossroads of environmental urgency and housing necessity, shipping container homes emerge as emblematic of a larger, vital movement towards sustainability. They are not merely homes, but symbols of possibility—a testament to what can be achieved when innovation meets intention. This narrative of transformation and possibility is what this book aims to explore. From the microcosm of individual homes to the macrocosm of global environmental impacts, the story of shipping container homes is a compelling chapter in the ongoing saga of sustainable living.

As the world grapples with the realities of climate change and resource scarcity, the principles underlying container homes—reuse, reduce, recycle—have never been more relevant. By choosing to live in these transformed spaces, individuals are not just making a statement about their personal living preferences but are also casting a vote for a type of development that values resourcefulness and environmental respect. This choice reflects a broader cultural shift towards sustainability that transcends economic and environmental benefits, touching the very core of how we perceive our roles as stewards of the earth.

The journey of sustainable living and the specific path of shipping container homes offer a vivid illustration of how necessity can drive innovation that, in turn, fosters greater societal change. This movement, grounded in the necessity to do more with less, challenges us to reconsider our lifestyles, our homes, and ultimately, the legacy we wish to leave on the planet. As this book unfolds, it will delve deeper into how these homes are designed, built, and inhabited, and how they represent a profound shift towards a sustainable future that is not just possible, but already in progress.

Chapter 1: Why Choose a Shipping Container Home?

The Eco-Friendly Appeal

In today's world, where the echoes of environmental change are more audible than ever, the choices we make about where and how we live can have profound impacts. The adoption of shipping container homes stands out as a robust response to the need for sustainable living practices, balancing human needs with environmental responsibility. These homes, crafted from retired shipping containers, have surged in popularity not just for their novelty, but for their significant ecological benefits. This movement toward container housing is not merely a trend; it is part of a larger, global shift towards sustainability that addresses numerous environmental concerns, from resource depletion to waste reduction.

One of the most compelling aspects of shipping container homes is their foundation in the principle of reuse. Millions of shipping containers are left idle across the globe, often seen as too expensive to send back empty to their origin. These containers, crafted from high-grade steel, are designed to withstand harsh conditions and heavy loads, making them structurally overqualified for simple storage but ideal for housing. By repurposing these containers for homes, we engage in an act of large-scale recycling, which drastically reduces the need for new construction materials like brick, wood, and mortar, which all carry heavy carbon footprints from production and transportation.

Moreover, the process of transforming these steel containers into livable spaces typically consumes far less energy compared to traditional construction methods. The basic structure already exists, which eliminates the initial phases of constructing a building from scratch and significantly reduces the amount of waste generated on construction sites. In addition, the inherent durability and weather resistance of these containers mean that they require less maintenance over time, contributing further to their sustainability.

Shipping container homes also promote a minimalist lifestyle that reduces resource use and environmental impact. The space constraints of a container encourage owners to prioritize essentials and downsize belongings, which in turn reduces consumption and waste. This minimalist approach extends to energy consumption. Smaller living spaces require less energy to heat and cool, which is not only eco-friendly but also economically beneficial. Many container homes are also equipped with green technologies such as solar panels, rainwater harvesting systems, and greywater recycling systems, enhancing their sustainability.

The placement of shipping container homes also frequently promotes a more sustainable lifestyle. These homes can be situated in diverse environments that traditional construction might disrupt more significantly. For instance, containers can be stacked and configured with minimal disturbance to the natural land, preserving the existing flora and fauna. Additionally, their robust nature allows them to be placed in locations prone to extreme weather, making use of sites that might otherwise be unsuitable for standard building techniques.

Beyond the environmental benefits, the architectural flexibility of container homes also speaks to the growing desire for unique, personalized living spaces that reflect individual values and aesthetics. Architects and builders can experiment with multiple configurations, cutting and combining containers in creative ways to form everything from compact studios to expansive multi-story residences. This adaptability not only recycles materials but also sparks innovation in the design field, challenging traditional architectural practices and encouraging new, sustainable ideas for modern living.

Furthermore, the community of shipping container home owners often shares a commitment to sustainable living that extends beyond their choice of housing. It is common for these individuals to engage in community-driven sustainability initiatives, such as communal gardens, shared renewable energy projects, and local recycling programs. This sense of community and shared purpose is pivotal, creating a culture of sustainability that influences others and magnifies the impact of choosing eco-friendly homes.

In essence, the eco-friendly appeal of shipping container homes is multi-faceted, touching on environmental, social, and economic levels. It represents a tangible action toward sustainable living that aligns with the values of reducing, reusing, and recycling, providing a practical and inspirational model of accommodation. As we continue to face global environmental challenges, the choice of a shipping container home is a profound step in aligning our living spaces with the principles of sustainability that are crucial for the long-term health of our planet. Thus, these homes do more than just provide shelter; they offer a forward-thinking approach to living that prioritizes the planet and its future, making a compelling case for why they are not just a housing option, but a lifestyle choice that resonates with the ethos of our times.

Financial Benefits and Savings

The financial allure of shipping container homes extends far beyond their initial cost-effectiveness. These homes present a groundbreaking approach to homeownership, especially appealing in an era where traditional housing costs are skyrocketing. This segment explores the myriad financial benefits that make shipping container homes an increasingly popular choice for those seeking affordable, yet stylish living solutions.

One of the primary financial advantages of shipping container homes lies in their affordability. The cost of a used shipping container can be remarkably low compared to traditional building materials. For many prospective homeowners, this significant reduction in initial outlay makes the dream of owning a home more attainable. Containers, once deemed surplus by shipping companies, are plentiful and relatively inexpensive, particularly when compared to the escalating prices of conventional construction materials like wood, brick, or stone.

Shipping container homes also benefit from reduced construction times. Since the primary structure of the home is already built, the amount of time required to convert a container into a livable space is considerably shorter than that needed to construct a house from scratch. This expedited building process cuts down on labor costs, which are often a substantial portion of a traditional home's overall budget. The ability to stack and join containers quickly also facilitates the creation of complex structures in a fraction of the time, further driving down construction costs. The inherent flexibility in the design and fabrication of container homes also contributes to their cost-effectiveness. Many shipping container modifications can be completed offsite in controlled environments, which maximizes efficiency and minimizes expensive onsite labor and equipment rental. Additionally, the DIY-friendly nature of container home construction appeals to hands-on individuals, who can choose to undertake much of the work themselves, thereby reducing professional labor costs even further.

Shipping container homes often embody energy efficiency, another avenue through which homeowners can realize financial savings. The ability to incorporate advanced insulation materials, innovative window placements, and green technologies like solar panels means that these homes can achieve significant reductions in energy use. This not only aligns with the eco-friendly appeal of container homes but also results in lower utility bills. Over time, the savings accrued from reduced energy consumption can substantially offset the initial investment in the home.

The robustness of shipping containers often allows for more straightforward foundation requirements compared to traditional structures, especially when dealing with durable surfaces or pre-existing concrete bases. This can lead to savings in both time and materials, as extensive excavation and site preparation are generally less necessary. Moreover, containers are designed to be secure and stable on varied terrains, which simplifies the logistics and reduces the costs associated with preparing land for construction.

As the popularity of container homes grows, so does their appeal in the housing market. These unique homes can attract significant interest, which may translate into higher resale values. The uniqueness of a well-designed container home, combined with the growing public interest in sustainable and innovative housing, can make these homes particularly appealing to a niche market, potentially driving up their resale price.

In some regions, shipping container homes are classified differently than traditional homes for tax and insurance purposes, potentially leading to lower property taxes or reduced insurance premiums. The specific classification can vary by location, but the potential financial benefits are significant. Additionally, the durability and security of steel containers can lead to lower insurance costs, as they are often more resistant to fire, wind, and termites than traditional homes.

The modular nature of shipping container homes simplifies future expansions or modifications. Homeowners looking to add more space can often do so at a lower cost and with less disruption than would be required with traditional construction. This ability to adapt and expand easily not only provides a financial benefit but also adds a level of future-proofing to the investment.

In conclusion, shipping container homes present a financially viable alternative to traditional housing. From the initial cost savings and reduced construction expenses to long-term energy efficiency and potential tax benefits, these homes offer a comprehensive array of financial advantages. The financial appeal, coupled with the environmental and design benefits, makes shipping container homes an intelligent choice for those looking to maximize their investment while achieving a sustainable, personalized living space.

Flexibility and Versatility in Design

When one thinks of a shipping container, the image that often comes to mind is that of a rectangular metal box, stark and functional, used for transporting goods across vast oceans. Yet, this seemingly mundane object has become a cornerstone in an architectural revolution that champions flexibility and versatility in home design. Shipping container homes, rising stars in the landscape of modern housing, offer a playground for innovation and creativity that traditional construction methods can seldom match.

The transformation of these robust steel boxes into comfortable, stylish homes is a testament to the adaptability inherent in container architecture. Architects and homeowners alike are drawn to the challenge of turning these industrial containers into living spaces that are not only functional but also aesthetically pleasing. The appeal lies in the almost limitless potential for customization and transformation that shipping containers offer. Whether one desires a compact and simple abode or a sprawling, multi-story residence, the modular nature of containers provides a framework that can be adapted to fulfill these diverse housing needs.

The primary allure of shipping container homes is the architectural freedom they afford. Each container is a prefabricated building block, ready to be modified, stacked, and interconnected. This modularity allows for designs that can range from single, minimalist cubes to complex configurations involving multiple containers aligned in innovative layouts. The strength of the steel frame also permits extensive cutouts for windows, doors, and even entire walls to be removed, creating open, airy spaces that defy the claustrophobic stereotypes often associated with container living.

This flexibility extends to the exterior aesthetics as well. Containers can be clad in a variety of materials, painted in any color, and finished to blend in with or stand out from their surroundings. The end result can be as understated or as bold as the owner desires. The inherent simplicity of the container is a blank canvas that invites imagination and personal expression, making each project uniquely reflective of its owner's tastes and lifestyle.

One of the most compelling aspects of shipping container homes is their ability to encourage sustainable and innovative use of space. In urban areas, where space is at a premium, the small footprint of these homes makes them ideal for fitting into tight spaces that traditional homes could not, such as narrow lots between existing buildings or unused backyards. Moreover, the possibility of vertical stacking harnesses the third dimension, offering solutions for increasing density without the sprawl.

Inside, the spatial design of container homes can be highly efficient. The constraints of the container's dimensions often lead to smarter, more innovative uses of space. Fold-out furniture, built-in storage, and multi-functional areas make these homes feel larger than their actual square footage. This efficiency is not just a matter of necessity but also a design philosophy that aligns with contemporary movements towards minimalism and eco-conscious living.

The needs of homeowners can change dramatically over time, and shipping container homes shine with their ability to adapt to these changes. Additional containers can be added to provide more space for growing families or new needs like home offices or studios. Conversely, as families shrink or needs change, sections of the home can be removed or repurposed without the extensive renovations required in traditional homes.

This adaptability makes container homes an excellent long-term investment. They can evolve as flexibly as the lives of their residents, accommodating shifts in family composition, lifestyle, or even location—if necessary, entire homes can be disassembled and relocated with far less effort and cost than traditional homes.

The design flexibility of shipping container homes also extends to their economic and ecological responsiveness. They can be outfitted with various green technologies, from solar panels and rainwater harvesting systems to high-efficiency insulation and HVAC systems. These additions are not afterthoughts but integral parts of the design process, tailored to each home's specific environmental and climatic conditions.

Furthermore, the ability to use up-cycled materials and integrate them with high-tech solutions allows for a unique balance between eco-friendliness and modern comforts. The result is a home that not only costs less to operate but also has a lower environmental impact throughout its lifecycle.

Finally, shipping container homes represent a form of cultural and creative expression that resonates with a diverse range of individuals. They embody the ideals of innovation, sustainability, and personalization, appealing to those who see their home not just as a place to live, but as an extension of their identity and values. Each container home tells a story, not just of its construction or design, but of the people who imagined and built it.

The flexibility and versatility of shipping container homes offer an exciting alternative to traditional housing. They provide a practical and artistic solution that meets the complex demands of modern living—economic efficiency, environmental sensitivity, and personal expression. As this trend continues to evolve, it challenges our preconceptions of what a home can be and encourages us to think creatively about how we define our spaces.

Chapter 2: Understanding the Basics

History and Evolution of Shipping Container Homes

The story of shipping container homes is one of innovation born out of necessity, and it serves as a fascinating chronicle in the broader narrative of human shelter. These homes, built from standardized reusable steel boxes, are more than just an architectural trend; they are a response to global economic, environmental, and social challenges. As we explore the history and evolution of shipping container homes, we trace the journey of these humble structures from mere cargo carriers to symbols of sustainable living.

The origins of shipping containers themselves date back to the mid-20th century, an invention credited to Malcolm McLean, an American trucking magnate and entrepreneur. In 1956, McLean introduced the shipping container to streamline the transportation of goods, thereby revolutionizing international trade. These containers were designed to be sturdy, weather-resistant, and capable of being easily transferred across different modes of transport—from ship to rail to truck—without unloading the cargo. This system, known as containerization, dramatically reduced shipping costs and improved efficiency, factors that contributed to the global spread of goods.

It wasn't until several decades later, however, that the potential of these containers to serve as more than just cargo holders was recognized. The first recorded instance of a shipping container being used for habitation was in the late 1980s. As urban populations grew and housing crises emerged, innovative architects and homeowners began to explore alternative living solutions that were both cost-effective and quick to deploy. Containers, often abundant and inexpensive due to the cost imbalance of shipping them back empty, presented an attractive solution.

The early adopters of container homes were often motivated by the practical benefits—low cost and availability—but they quickly discovered the additional advantage of durability. Containers, built to endure harsh marine environments and heavy loads, could easily withstand terrestrial weather challenges. This durability was paired with an ease of modification that intrigued the architectural community. The steel structure could be cut and reconfigured without compromising its integrity, allowing for customizations that traditional building materials did not permit as easily.

The movement gained momentum in the early 2000s as environmental concerns became more pronounced. The concept of upcycling—transforming waste materials into new products of higher quality or value—found a perfect match in shipping container architecture. Each reused container was a step away from the resource-intensive process of new construction, significantly reducing the carbon footprint of building a home.

Architects and builders began to push the boundaries of what could be achieved with these metallic boxes. Pioneering projects showcased containers as versatile building blocks that could be stacked, sliced, and shifted to create everything from single-room homes to expansive multi-level residences. The visual appeal of these structures also evolved; what began as utilitarian constructions soon included stylish, modern designs that challenged any preconceived notions about the aesthetics of eco-friendly housing.

As container homes rose in popularity, they started appearing not just in remote or experimental settings but in urban centers. Cities facing housing shortages and high construction costs saw container homes as a viable alternative. They could be built quickly in response to immediate needs, such as disaster relief housing or temporary accommodations for events or seasonal workers. Furthermore, their modular nature allowed for communities to be formed with shared amenities, fostering a sense of neighborhood in settings that required rapid development.

Internationally, shipping container homes have been embraced in a variety of cultural contexts. In places like Amsterdam and London, they are used as student housing solutions. In Australia, they serve as both beach homes and bushfire-resistant dwellings. Meanwhile, in parts of Africa and Asia, containers provide a robust framework for developing housing quickly and affordably in growing urban areas.

Today, the narrative of shipping container homes is intertwined with the global movement towards sustainability. They are seen not only as a smart financial choice but as part of a larger ethos of responsible, conscious living. Innovations continue to emerge, with technology playing a significant role in making these homes more comfortable, livable, and energy-efficient.

This historical perspective highlights the adaptability and resilience not just of the materials but of the human spirit in its quest for sustainability. From their utilitarian origins to their present status as icons of eco-friendly architecture, shipping container homes represent a critical evolution in the way we think about space, resources, and our environmental impact. As we move forward, they symbolize our ongoing commitment to building smarter, living better, and treading more lightly on the earth.

Types of Shipping Containers and Their Uses

In the evolving narrative of shipping container homes, understanding the types of containers available and their specific characteristics is crucial. These containers, initially designed for transporting goods across oceans, have morphed into versatile building blocks for innovative homes. Each type offers unique features and possibilities for residential conversion, making the selection process a foundational step in designing a container home.

Standard Dry Storage Containers

The most common type of shipping container used in building homes is the standard dry storage container, typically made from weathering steel known as COR-TEN steel, which is known for its rust resistance and durability. These containers come in a variety of sizes, with the most popular being the 20-foot and 40-foot lengths. Both sizes are about eight feet wide and eight and a half feet tall, providing a compact yet spacious environment for a small home or an individual living space.

The 20-foot container, offering about 160 square feet of space, is often chosen by those looking to create smaller, modular homes or single residential units. Its size makes it easily transportable and perfect for a minimalist lifestyle or a starter home for individuals and couples. On the other hand, the 40-foot container doubles the living space to approximately 320 square feet, making it suitable for a more extensive living area or multiple rooms. This type is favored for family homes or larger projects where space and flexibility are desired.

High-Cube Containers

High-cube containers are similar to standard containers in length and width but have an extra foot in height, standing at about nine and a half feet tall. This additional space makes high-cube containers particularly appealing for residential use, as the extra ceiling height can help the home feel more spacious and open. This feature is especially beneficial when incorporating insulation and interior finishes, which can decrease the ceiling height in standard containers. The added volume accommodates more creative design elements, such as lofted areas or high storage solutions, enhancing the functionality and aesthetic appeal of the living space.

Open Top, Open Side, and Flat Rack Containers

While less common in residential conversions, other types of containers like open top, open side, and flat rack also provide unique opportunities. Open top containers, with removable tops typically covered by a tarpaulin, allow for the insertion of large items that might not fit through the doors, which can be an advantage during the initial phase of converting the container into a home.

Open side containers feature doors that open on one or both long sides in addition to the end doors, greatly enhancing accessibility and allowing for more natural light and airflow throughout the home. This type can be particularly effective for creating a dynamic living space that opens up to an outdoor environment, ideal for locations with favorable climates.

Flat rack containers, which have collapsible sides and are primarily used for heavy or oversized cargo, offer a robust framework for constructing larger spaces when locked in place. These are less frequently used for full homes but can be excellent for additions like decks, patios, or even bridges connecting multiple containers.

Refrigerated ISO Containers

Refrigerated containers, or reefers, are insulated units designed to transport perishable goods. While their insulation makes them initially more expensive and less commonly used in residential projects, they offer significant advantages in terms of energy efficiency. The built-in insulation can reduce the need and cost for additional thermal lining, making them an excellent choice for homes in both extremely cold and hot climates.

Specialized Containers

A variety of specialized containers are also available, including those designed for specific industrial needs, which can be repurposed for residential use. For example, tank containers, which are used to transport liquid materials, can be converted into unique architectural features like swimming pools or hot tubs.

Practical Considerations in Selecting Containers

Choosing the right type of container for a home project involves considering several practical aspects, such as the intended design, budget, and the climatic conditions of the area. It is crucial to inspect containers for structural integrity, especially if they are used, as they may have been subjected to considerable wear and tear during their shipping careers. Furthermore, understanding the limitations and possibilities of each container type helps in planning effective layouts and ensuring that the home can be both comfortable and aesthetically pleasing.

Versatility in Use

Beyond residential uses, shipping containers have been employed in various other capacities that highlight their versatility. These include commercial structures like pop-up shops and offices, educational facilities, emergency hospitals, and artistic installations. Each of these uses underscores the adaptability of shipping containers, transforming them from simple cargo carriers into dynamic spaces that cater to a wide range of human activities.

In conclusion, the variety of shipping containers available today provides a rich palette for architects and builders to draw from. The decision to use a particular type depends on a complex interplay of factors including functionality, design preferences, budget constraints, and environmental considerations. Understanding these options allows potential builders and homeowners to make informed decisions, paving the way for innovative and personalized container homes that push the boundaries of traditional architecture.

Common Misconceptions Debunked

As the popularity of shipping container homes continues to rise, so too do the misconceptions surrounding this innovative housing solution. Misunderstandings can range from questions about structural integrity to doubts about comfort and livability. Addressing these common misconceptions is crucial not only for prospective builders and homeowners but also for the broader acceptance of shipping container homes as a viable and sustainable housing option.

One of the most prevalent myths about shipping container homes is that they lack the durability and safety of traditional homes. Critics often cite the containers' original purpose—transporting goods across long distances in harsh weather conditions—as a potential drawback, suggesting that they might not be suitable for permanent housing. However, the reality is quite the contrary.

Shipping containers are designed to endure extreme conditions on sea voyages, making them incredibly sturdy. Constructed from high-grade steel, they are capable of withstanding tremendous pressure and weight, attributes that actually enhance their suitability as building materials. When properly maintained and treated, containers can resist corrosion and serve as robust, secure homes for decades. Additionally, their ability to be easily modified allows for the installation of modern safety features, such as smoke detectors, insulation, and weatherproofing enhancements that meet or even exceed building codes.

Another common misconception is that container homes are challenging to keep warm in the winter and cool in the summer due to their metal construction. While untreated metal can indeed transmit temperature changes rapidly, proper insulation solves this problem. Modern insulation techniques can make a container home just as energy-efficient as any traditionally built house. Options like spray foam insulation, which adheres directly to the interior walls of the container, create an airtight and watertight seal that significantly improves thermal performance. Moreover, installing innovative green roofing systems or utilizing strategic window placements can enhance energy efficiency, making container homes comfortable in any climate.

To the uninitiated, it might seem that the uniform shape and size of shipping containers would restrict design creativity. Yet, the truth is that these building blocks can be combined and configured in countless ways. Architects and designers have demonstrated time and again that with imaginative planning, containers can be transformed into luxurious, multi-story residences, cozy cottages, or anything in between. The inherent modularity of containers also allows for expansive design flexibility—units can be stacked and aligned to create various shapes and layouts that challenge traditional architectural forms.

Interior design in container homes also offers vast potential. With the right approach, the interiors can be finished to reflect any style, from ultra-modern minimalism to warm, rustic charm. The long, narrow shape of a typical shipping container is comparable to many modern micro-apartment designs in cities around the world, proving that space constraints can inspire rather than inhibit creativity.

A persistent myth is that obtaining building permits for container homes is more difficult than for traditional homes. This misconception stems from unfamiliarity with the concept among local planning offices rather than from any inherent legal complexities associated with container homes. As container homes become more common, many urban areas are updating their building codes to accommodate this type of construction. Educating local authorities and demonstrating compliance with all relevant codes are key to navigating this process smoothly. Additionally, many architects and builders specialize in container homes and can help guide homeowners through the permitting process.

Some critics argue that the eco-friendly label of container homes is overstated, pointing out that the energy used to modify containers might negate the benefits of recycling them. However, when compared to the environmental impact of traditional construction—which involves extensive mining, logging, and other resource-extractive activities—the reuse of shipping containers is considerably less disruptive to the planet. Furthermore, container homes often incorporate other sustainable practices such as the use of renewable energy sources and water-saving fixtures, reinforcing their status as a green housing option.

Addressing these misconceptions not only clarifies the reality of living in a shipping container home but also highlights the innovative and sustainable potential of this housing alternative. As more people discover the truth about container homes, the interest and investment in this type of housing are likely to grow, paving the way for a more sustainable and creative approach to residential construction. By dispelling myths and spreading knowledge, we can foster a more accurate understanding of what it means to live in, and build, a shipping container home.

Chapter 3: Planning Your Container Home Project

Intro

Embarking on the construction of a shipping container home is a bold, creative endeavor that embodies a modern approach to sustainable living and unique architectural design. This chapter, "Planning Your Container Home Project," is designed to guide you through the intricate process of turning your vision for a container home into a tangible, livable reality. It addresses the foundational steps necessary for successful planning and execution, covering everything from setting realistic goals and budgets to designing with efficiency and sustainability in mind, and navigating the often-complex terrain of zoning laws and building codes.

The process of building a container home requires more than just an innovative idea and a plot of land. It demands a detailed understanding of what is feasible both financially and practically, considering the limits and opportunities presented by local regulations and the physical characteristics of shipping containers themselves. This chapter aims to equip you with the essential knowledge and tools to meticulously plan your project, ensuring that you make informed decisions that align with your long-term living needs and environmental values.

We begin by examining how to set realistic goals and budgets, an essential step that lays the groundwork for all future decisions in the home construction process. This section will help you understand the various cost factors involved in container home construction, from acquiring the containers themselves to modifying them and complying with local building standards. It will also explore how to allocate resources wisely, ensuring that you can achieve your dream home without compromising financial stability.

Next, we delve into designing for efficiency and sustainability, crucial for those who wish to minimize their ecological footprint while maximizing the functionality and aesthetic appeal of their home. This section provides insights into the selection of materials, energy use strategies, and design principles that can enhance the livability and sustainability of your container home. It highlights innovative techniques and technologies that make container homes not only more environmentally friendly but also cost-effective in the long run.

Lastly, the chapter addresses the critical aspect of navigating zoning laws and building codes, which can often be the most daunting part of planning a container home project. Understanding these regulations is vital to ensure that your home is not only safe and structurally sound but also legally compliant. This section offers practical advice on how to engage with local authorities, secure the necessary permits, and meet or exceed the required building standards.

By the end of this chapter, you will be equipped with a comprehensive understanding of the planning stages of building a shipping container home. This knowledge will empower you to move forward confidently, making decisions that are informed, strategic, and aligned with your vision of creating a unique, sustainable home.

Setting Realistic Goals and Budgets

Undertake the journey to build a shipping container home is an adventure that marries creativity with practicality. It's an endeavor that requires a clear vision, supported by well-defined goals and a meticulously planned budget. As potential builders contemplate this exciting venture, understanding the full scope of what's involved is critical to turning their dream home into a reality without unforeseen setbacks or financial strain.

Before diving into any construction project, especially one as unique as a shipping container home, it's essential to define what you aim to achieve. Are you building a compact bachelor pad, a family residence, or perhaps a multi-container retreat? Your goals will determine the scale and complexity of the project and have a direct impact on your budgeting. It's crucial at this stage to assess your needs versus wants, establishing priorities that will guide the entire project from conception to completion.

One of the first steps in setting a realistic budget is understanding the costs associated with purchasing and converting shipping containers into livable spaces. These costs can vary significantly depending on several factors including the condition of the containers, the complexity of the designs, and the location of the build. Initial expenses typically include the purchase price of the containers, delivery charges, and the costs of any permits or legal fees required to begin construction.

Beyond these initial costs, prospective builders must also consider the expenses related to site preparation, foundation setting, insulation, interior finishing, utilities installation, and exterior modifications. Each of these stages can vary in cost depending on the choices made. For example, opting for high-end finishes or advanced eco-friendly technologies can drive up expenses considerably.

The complexity of your container home's design will significantly affect your budget. Complex designs that require extensive structural modifications, such as removing entire sides of a container or adding large glass panels, can escalate costs not only due to the materials but also because of the need for skilled labor. Similarly, the choice of materials, both for the interior and exterior, can impact the budget. Eco-friendly or premium materials typically come at a higher cost but can reduce long-term expenses through energy savings and durability. Labor costs can be one of the most significant expenses in building a container home, particularly if you rely on professionals for most of the construction work. An alternative to reduce this cost is taking on a DIY approach to some aspects of the build. However, it's important to realistically assess your skills and the time you can dedicate to the project. Overestimating your capabilities or available time can lead to delays and potentially higher costs if professionals are needed to correct mistakes.

A wise budget for any construction project includes a contingency fund, typically 10-20% of the total budget, to cover unexpected costs that arise during construction. These might include issues with the land, unforeseen legalities, or adjustments in design necessitated by practical considerations discovered during the build. Having this buffer can be the difference between completing your project or facing stressful financial hurdles.

When setting your budget, it's also beneficial to consider the long-term financial implications of owning a container home. This includes maintenance costs, potential energy savings from sustainable designs, and the resale value of the home. Efficient planning can lead to significant reductions in ongoing costs, making the container home not just an affordable housing solution upfront but also a cost-effective choice in the long term.

Financing a container home can be challenging as traditional mortgage options may not always be available. It's crucial to explore various financing routes early in the planning stage. Options may include personal loans, construction loans, or specialty financing through companies that recognize the unique nature of container homes. Understanding these options and securing financing before beginning construction can ensure that the project progresses smoothly without financial interruptions.

Finally, maintaining flexibility in both your goals and your budget is essential. Regularly review and adjust your plans as the project progresses and as you gain a clearer insight into the practicalities of building a container home. This adaptiveness can help manage costs effectively, allowing adjustments to be made before budgets are overstretched.

Setting realistic goals and a detailed budget for a container home project demands thorough planning and a clear understanding of the costs involved. By carefully considering each element of the project from the outset, prospective homeowners can ensure that their venture into the world of container homes is both successful and satisfying, turning their architectural dreams into a durable and delightful reality.

Design of a small, modern Nordic-style eco-friendly container house

Designing for Efficiency and Sustainability

Creating a shipping container home is not just about transforming a steel box into a living space; it's about doing so in a way that promotes efficiency and sustainability. This approach not only ensures the home is comfortable and functional but also minimizes its impact on the environment. Achieving this requires thoughtful design that integrates advanced building techniques, innovative materials, and smart technology—all tailored to meet the unique challenges and opportunities presented by container architecture.

Sustainable design in the context of shipping container homes means creating spaces that are energy-efficient, environmentally friendly, and economically viable. It involves considerations like material sourcing, energy consumption, and the long-term impact of the construction processes. Effective sustainable design enhances the inhabitants' quality of life by optimizing the indoor environment and reducing operational costs through energy savings.

One of the first steps in designing an efficient container home is choosing the right placement and orientation. This decision can significantly impact the home's thermal performance and energy consumption. For example, positioning a home to take advantage of natural light can reduce reliance on artificial lighting. Similarly, considering prevailing winds can help in naturally cooling the home, reducing the need for air conditioning. The orientation can also affect the effectiveness of solar panels, which are a common feature in container homes due to their flat roofs that offer an ideal mounting surface.

Passive design is a key strategy in making container homes more sustainable. This approach uses the climate to maintain a comfortable temperature range in the home, minimizing the use of mechanical heating or cooling systems. Key elements include insulation, thermal mass, and ventilation. Proper insulation is critical in container homes, as metal conducts heat rapidly. Advanced insulation techniques not only preserve interior temperatures but also prevent condensation—a common issue in metal structures.

Thermal mass, which refers to materials that can absorb and store heat energy, is another important aspect. In container homes, this can be achieved through internal elements like concrete floors or walls, which help regulate the home's temperature by absorbing heat during the day and releasing it at night.

Natural ventilation should be strategically designed to allow air to flow through the home efficiently, reducing the need for electric fans or air conditioners. This can be achieved through the placement of windows, doors, and vents to facilitate cross-ventilation.

Selecting the right materials is crucial for the sustainability of container homes. This includes choosing materials that are recycled, recyclable, or from renewable sources. For instance, bamboo flooring is a popular option due to its renewability, durability, and aesthetic appeal. Low-VOC (volatile organic compounds) paints and finishes can improve indoor air quality by reducing toxins.

In addition to materials, integrating green technologies is vital. Solar panels, rainwater harvesting systems, and greywater recycling systems can all enhance a home's sustainability. These technologies not only reduce the home's environmental footprint but also lower utility bills.

The integration of smart home technology can further enhance the efficiency of a container home. Smart thermostats, LED lighting systems, and energy-efficient appliances can all be controlled via smartphones, optimizing energy use throughout the home. These technologies provide residents with the tools to monitor and manage their energy consumption actively, leading to more sustainable living habits.

Another aspect of sustainable design is building for the future. This means considering not just the current needs but also how they might change over time. Designing a container home that can be easily modified or expanded can prevent future demolition and reconstruction, reducing waste and costs. For example, designing plumbing and electrical systems that can be easily accessed and modified can allow for changes in the home's layout or the addition of new technologies without significant disruptions.

Finally, sustainable design also involves considering the broader impact of the construction project. This includes the effects on the local community and the environment. Using local labor and materials can reduce the carbon footprint associated with transporting goods, while also supporting the local economy.
Designing for efficiency and sustainability in container homes is about making informed choices that balance comfort, functionality, and environmental responsibility. By embracing innovative design strategies and technologies, builders can create container homes that are not only energy-efficient and beautiful but also have a positive impact on the environment and their inhabitants' well-being. These homes stand as a testament to the possibilities of sustainable architecture, offering a blueprint for responsible residential design in the 21st century.

Navigating Zoning Laws and Building Codes

Embarking on the construction of a shipping container home involves more than just design and assembly; it requires a deep understanding and adherence to zoning laws and building codes. These regulations are crucial as they ensure safety, sustainability, and compatibility with the community. Navigating these laws can be one of the most challenging aspects of building a container home, especially for those new to the world of construction and real estate development.

Understanding Zoning Laws

Zoning laws are local and regional legal stipulations that govern land use. They dictate what types of structures can be built in certain areas, how land can be used, and specific requirements for buildings, including their size, placement, and function. For potential container homeowners, understanding these laws is the first step in planning the construction process.

Zoning laws can vary significantly from one locality to another, which means that what is permissible in one area might be prohibited in another. For instance, some regions might embrace the innovative nature of container homes, promoting them as part of eco-friendly development initiatives, while others may have restrictions due to historical preservation, aesthetic standards, or urban planning strategies. Therefore, the first step for any container home builder should be to contact local zoning offices to gather all pertinent information regarding permissible land use.

Engaging with Local Planning Departments

Engaging proactively with local planning departments is essential. This engagement can provide insights into specific zoning restrictions and what exceptions or variances might be possible. Planning departments can offer guidance on the necessary steps to obtain variances, which are permissions granted to deviate from certain zoning requirements. For container home builders, obtaining a variance might be necessary if their project does not strictly comply with local zoning laws.

Builders should prepare detailed plans and rationales for their projects when approaching local authorities. This preparation shows a commitment to compliance and community standards, which can facilitate smoother interactions with zoning boards and other regulatory bodies. In some cases, presenting the environmental benefits and innovative design aspects of container homes can help in gaining favorable responses.

Deciphering Building Codes

Building codes, unlike zoning laws, specifically focus on the construction standards of buildings. They ensure that structures are safe and accessible and meet national or international safety standards. Building codes cover a wide range of construction and design aspects, including structural integrity, fire safety, electrical systems, plumbing, ventilation, and accessibility.

Container homes, by their very nature, may present unique challenges in meeting these standards. For example, modifying the structure of a shipping container, such as cutting through steel walls to create windows or doors, can affect its structural integrity. Therefore, it's crucial to work with engineers or architects who have experience with container modifications to ensure that any alterations comply with building codes.

Integrating Fire Safety and Insulation Standards

One of the critical areas where building codes come heavily into play is in fire safety and insulation. Containers are primarily made of steel, which is a good conductor of heat. Adequate insulation is necessary not only for comfort but also for safety, preventing potential fire hazards and ensuring energy efficiency. Building codes may dictate specific types of insulation materials and installation methods. Compliance with these requirements is crucial not only for gaining building permits but also for ensuring the long-term safety and livability of the home.

Accessibility and Environmental Compliance

Another important consideration is ensuring that the container home meets accessibility standards as outlined in building codes, such as the Americans with Disabilities Act (ADA) in the United States. These standards are designed to ensure that homes are accessible to people with disabilities, featuring modifications like wider doorways, accessible bathrooms, and ramps.

Additionally, environmental regulations may affect certain aspects of the construction process, including waste disposal, water use, and the use of certain materials. Compliance with environmental standards not only supports sustainability but can also affect the overall approval and success of the project.

Building a Support Network

Since navigating zoning laws and building codes can be complex, building a support network of knowledgeable professionals such as real estate attorneys, experienced contractors, and architects can prove invaluable. These professionals can offer advice, identify potential legal issues, and help streamline the approval process.

Successfully navigating zoning laws and building codes is fundamental to the construction of a shipping container home. By thoroughly understanding and adhering to these regulations, builders can ensure that their container home project is not only innovative and environmentally friendly but also safe, legal, and welcomed within their community. This process, while sometimes daunting, is a crucial investment in the future success and sustainability of the container home.

As we conclude this chapter on planning your container home project, it's clear that the journey from concept to completion involves a series of detailed and deliberate steps. Each phase of planning and execution brings its own set of challenges and rewards, necessitating a balanced approach to creativity, practicality, and diligence. The insights provided here are intended to pave a path that leads to the successful realization of your container home, ensuring that every aspect of the project aligns with both your personal aspirations and regulatory requirements.

The initial step of setting realistic goals and budgets is more than just a preliminary task; it is a crucial strategy that influences every other decision in the construction process. By establishing a clear and feasible financial plan, you position yourself to navigate the ups and downs of home building with resilience and foresight. This foundational step ensures that the project remains viable and sustainable from start to finish, reflecting your financial limits and lifestyle needs.

Designing for efficiency and sustainability has underscored the importance of incorporating eco-friendly practices and materials into your container home. This approach not only supports environmental conservation but also enhances the quality of life within the home through improved energy efficiency and reduced operational costs.

The creative adaptation of shipping containers for housing purposes showcases the potential for innovative, sustainable living solutions that can be both beautiful and functional.

Navigating zoning laws and building codes is perhaps the most complex aspect of planning your container home. This stage demands thorough research, proactive engagement with local authorities, and a deep understanding of the legal landscape. The knowledge gained through this process is invaluable, ensuring that your home meets all necessary safety and compliance standards, which is essential for securing the longevity and legality of your construction.

The journey to create a shipping container home is a testament to the power of innovative thinking in contemporary architecture. It challenges conventional norms and pushes the boundaries of traditional home building. As you move forward, armed with the knowledge and strategies outlined in this chapter, you are well-prepared to turn your vision into reality. Your future container home is not just a structure but a personal statement of creativity, sustainability, and resilience.

This chapter has laid the groundwork for your project, providing the tools and knowledge needed to navigate the complex yet rewarding process of building a shipping container home. With careful planning, a commitment to sustainability, and a thorough understanding of legal requirements, your container home will stand as a beacon of innovative, environmentally conscious living.

Chapter 4: Acquiring Your Containers

Where to Find and How to Choose the Right Containers

Undertaking a container home project starts with one fundamental task: finding and selecting the right shipping containers that will form the building blocks of your new home. This stage is more than just a logistical step; it is a critical decision that affects the sustainability, quality, and overall success of your project. To navigate this phase effectively, you must understand where to look for containers, how to assess their suitability, and what factors influence their condition and price.

Shipping containers are abundant, but finding the right type at a reasonable price requires some know-how and strategy. The most common sources for acquiring containers include:

1. **Shipping Yards and Depots**: Often the first stop in searching for containers, these facilities store the containers used for international shipping. Purchasing directly from shipping yards can be cost-effective, especially if you can inspect the containers in person.

2. **Container Retailers and Dealers**: Specialized dealers often offer a variety of containers, including new and used ones. They can provide valuable services such as delivery and customization. Retailers typically have a good understanding of the types of containers that are best suited for home construction.

3. **Online Marketplaces**: Websites like Craigslist, eBay, and specialized online portals for container sales can offer competitive prices and a wide selection of containers. Buying containers online, however, usually prevents a pre-purchase physical inspection, which can be a gamble regarding the container's condition.

4. **Local Classifieds and Auctions**: Sometimes businesses or individuals sell containers through local classified ads or at auction. These can be excellent opportunities to find deals, especially if the containers are no longer deemed viable for commercial shipping but are still structurally sound.

5. **Custom Container Companies**: Some companies specialize in modifying shipping containers for specific uses, including homes. These companies can be a one-stop shop, providing both the container and modification services tailored to building regulations and personal preferences.

Once you have located potential containers, the next step is assessing their quality and suitability for your project. Key factors to consider include:

- ***Age and Condition***: Older containers may have more wear and tear, which could include rust, dents, and damaged flooring. The container's age can also affect its structural integrity, which is critical in a home setting.

- ***Previous Usage***: Understanding what the container was used for previously can help assess potential contamination from hazardous materials, which is especially important if the container will form part of a living space.

- ***Size and Type***: Containers come in various sizes and types, such as standard, high-cube, and refrigerated models. Choosing the right size and type depends on your design requirements and the intended use of each space within your home.

- ***Certifications***: Some containers are certified for specific uses, such as "one-trip" containers, which have only been used once and are typically in better condition than those that have been in long-term shipping use.

If possible, conduct a physical inspection of the container before purchase. This inspection is crucial for identifying issues that might not be apparent in photos or descriptions, such as:

- ***Structural Integrity***: Check for significant rust, especially at the corners and along the bottom edges where moisture can cause substantial damage. Ensure the doors function properly and the seals are intact.

- ***Interior Condition***: Inspect the interior for signs of mold, undesirable odors, or chemical residues, which could pose health risks or additional cleaning and maintenance costs.

- ***Modifications***: Sometimes containers are sold with modifications already made, which might be beneficial or detrimental depending on their quality and your needs.

Choosing the right container is about balancing cost with quality. Opting for the cheapest option might result in higher refurbishment expenses if the container requires extensive repairs. Conversely, investing in a container in better condition might streamline the preparation and modification process, ultimately saving money and time.

In summary, finding and choosing the right shipping containers for your home project requires a diligent approach to sourcing, a thorough inspection to assess their condition, and a strategic selection to match your specific design and living needs. This initial step lays a solid foundation for the transformation of these robust steel boxes into a comfortable, sustainable home. By starting with the right materials, you ensure a smoother process throughout the subsequent phases of your container home construction.

Inspecting Containers for Quality and Safety

Choosing the right shipping container for a home project goes beyond just picking one off a lot. The integrity of your future home depends heavily on the quality and condition of these steel boxes, making thorough inspections an indispensable part of the purchasing process. A detailed examination not only ensures safety and durability but also helps avoid costly repairs that could escalate your project's budget unexpectedly.

The life of a shipping container before it retires into your home can be harsh. These containers travel thousands of miles, often through extreme weather conditions, and can carry everything from harmless household goods to hazardous chemicals. Each has a story, and sometimes that story can leave a container worn or even structurally compromised. Detailed inspections help filter out containers that might look adequate on the outside but have underlying issues that could compromise your home's integrity and safety.

The first and most critical aspect of the inspection is to assess the container's structural integrity. This involves a close examination of several key components:

- ***Corners and Edges***: These areas bear much of the structural load and are prone to damage during handling and transport. Check for any significant dents or deformations that might affect the structural strength.

- ***Walls and Roof***: Inspect the walls and roof for rust patches, holes, or severe dents. Minor surface rust can be treated, but deep corrosion may indicate that the steel has been compromised.

- ***Flooring***: Most container floors are made from marine-grade plywood, which is durable and resistant to rot. However, it's essential to check for signs of water damage or heavy wear, which could suggest previous leaks or exposure to moisture.

- ***Door Mechanism***: The doors of a container must close tightly and securely. Inspect the door hinges and locking mechanisms for rust or damage, as these could affect the seal and security of the container.

Beyond structural integrity, safety is another paramount concern when inspecting a container for residential use. Here are key safety aspects to consider:

- ***Chemical Contamination***: Containers used for shipping goods are often treated with pesticides or may have transported toxic substances. It's crucial to determine the container's cargo history and inspect for any residues that might pose health risks. Sometimes, a professional cleaning or replacing the plywood floor is necessary to ensure the container is safe for residential use.

- ***Lead-Based Paint***: Older containers may have been painted with lead-based paints, which pose serious health risks. Check the paint condition and consider encapsulation or removal by professionals if lead paint is suspected.

- ***Asbestos***: In some older containers, asbestos might be present in gaskets or sealants. This requires professional assessment and removal to prevent any exposure to asbestos fibers.

While much of the inspection can be performed visually, there are instances where more advanced techniques might be necessary:

- ***Thermal Imaging***: This can be used to detect areas of missing or inadequate insulation, as well as moisture intrusion that is not visible to the naked eye.

- ***Ultrasonic Testing***: For a more detailed assessment of the steel's condition, ultrasonic testing can detect internal corrosion and thickness reduction, which might not be apparent during a visual inspection.

- ***Pressure Testing***: To ensure the container is airtight and watertight, conducting a pressure test can be beneficial, especially if the container will be modified extensively.

Documenting every step of the inspection process is crucial, not only to maintain records of what was inspected and what issues were found but also to have evidence of the container's condition at the time of purchase. This documentation can be helpful for insurance purposes, future modifications, or if legal issues arise.

While buyers can perform many inspection tasks themselves, hiring a professional inspector or a structural engineer is advisable, especially for those unfamiliar with container structures. These professionals can provide more detailed insights and help ensure that the containers you choose are fit for their intended purpose as building blocks of your home.

In conclusion, inspecting containers for quality and safety is a meticulous but necessary process that underpins the success of a container home project. By taking the time to thoroughly inspect potential containers, you ensure that your foundation is solid, safe, and ready for transformation into a dream home. This proactive approach minimizes unexpected challenges and ensures that the container home will be a secure and lasting abode.

Logistics of Delivery and Placement

The logistics of delivering and placing shipping containers for a home project involve a complex array of considerations. This step in your container home journey is critical, as it not only involves the physical transport of large, heavy steel boxes but also the strategic placement to ensure the future stability and accessibility of your home. The process demands meticulous planning, coordination with various professionals, and an understanding of the technical challenges that may arise. Addressing these logistical elements effectively ensures a smooth transition from a mere plot of land to the foundational setup of your new container home.

Before the containers even arrive on your site, significant preparation is necessary to ensure a smooth delivery process:

- ***Site Access Evaluation***: Assess the accessibility of your site. Consider factors like road width, bridge clearance, and the strength of local road networks. The route to your site must be able to accommodate large trucks, potentially with heavy loads. Check for any local restrictions or permits required for transporting oversized loads on public roads.

- ***Site Preparation***: The land where the containers will be placed needs to be prepared. This involves clearing debris, leveling the ground, and sometimes laying a temporary or permanent foundation. The type of foundation depends on the soil type, the weight of the container, and local building codes.

- ***Scheduling Deliveries***: Coordination with the delivery company is crucial. The timing of the delivery should align with your project schedule, considering factors like weather conditions, availability of construction crews, or other scheduled site work.

The method of delivery might vary based on the size of the containers, the number and type of containers being delivered, and the specifics of your site:

- ***Tilt Bed Truck***: One of the most common methods for delivering containers. The truck tilts to slide the container off the bed, requiring a good amount of clear space in front of the delivery site for the truck to maneuver and for the container to land.

- ***Flatbed Truck and Crane***: For sites with limited space or where multiple containers need to be stacked, a flatbed truck may be used in conjunction with a crane. The crane lifts the containers off the flatbed and places them precisely on the foundation. This method is more expensive but offers greater placement precision.

- ***Side Loader***: This is effective for tight spaces where the container needs to be loaded from the side. It uses a specialized truck that lifts the container from the side and sets it down parallel to the vehicle.

Physical constraints of the site can greatly influence the delivery and placement process:

- ***Space Constraints***: Limited space may require additional machinery like cranes or specialized lifting equipment. Urban settings often pose unique challenges, such as overhead wires or nearby structures, which need to be considered.

- ***Ground Conditions***: Soft ground may need reinforcement to support heavy trucks and cranes. In some cases, temporary roadways might be installed to prevent vehicles from becoming bogged down.

- ***Container Placement Strategy***: The final placement of the container should be planned in advance, considering how each container fits within the overall design of the home. The orientation of containers can affect future functionality and aesthetics, including factors like natural light, privacy, and wind direction.

Once the containers are delivered and placed, the initial setup tasks begin:

- ***Alignment and Leveling***: Containers need to be perfectly aligned and leveled to ensure structural integrity, especially if they will be stacked or joined. This might involve adjusting the containers on their foundations or using jacks to level them.

- ***Securing Containers***: Containers should be securely fastened to their foundations. This is crucial not only for stability but also for safety, particularly in areas susceptible to high winds or seismic activity.

- ***Weatherproofing***: Soon after placement, containers should be weatherproofed to prevent any potential damage. This includes sealing any gaps and protecting exposed metal to prevent rust.

- ***Utilities and Modifications***: Early steps should be taken to plan for utilities like water, electricity, and sewer connections, as well as structural modifications such as cutting openings for windows and doors.

Navigating the logistics of container delivery and placement is a complex but achievable task. It requires careful planning, coordination, and flexibility. Overcoming these logistical hurdles sets a solid foundation for the rest of the construction process, moving you one step closer to realizing your vision of a container home. This stage not only represents the physical manifestation of your project but also symbolizes the tangible beginning of building not just a structure, but a home.

Chapter 5: The Design Phase

Principles of Eco-Friendly Design

The design phase of creating a container home is not just about aesthetics and functionality; it's fundamentally about integrating principles of eco-friendly design that align with broader goals of sustainability and environmental stewardship. As the world becomes increasingly aware of the impacts of climate change and resource depletion, the imperative to design homes that are energy-efficient, minimize waste, and utilize environmentally friendly materials has never been more critical. In this context, container homes present a unique opportunity to redefine what it means to live sustainably. This section explores the foundational principles of eco-friendly design as applied to the construction of container homes, emphasizing how these structures can contribute to a greener future.

Eco-friendly design, also known as sustainable design, aims to reduce the environmental impact of buildings through thoughtful and efficient use of resources, energy, and space.

It involves a holistic approach that considers every phase of a building's life cycle, including design, construction, operation, and demolition. For container homes, eco-friendly design starts with the very ethos of reusing shipping containers as a primary building material, thereby diverting them from scrap yards and reducing the demand for new raw materials.

The use of shipping containers themselves is a form of resource efficiency, a key principle of eco-friendly design. These containers are typically discarded after their shipping life, which can be as little as 5-10 years. Reusing these structures saves a significant amount of steel and, by extension, the energy and raw materials required to produce new steel products. However, resource efficiency extends beyond just reusing containers. It also involves:

- ***Sourcing Local Materials***: Whenever additional materials are needed, sourcing locally reduces transportation emissions and supports local economies. For instance, using locally sourced timber for framing or finishes not only minimizes the carbon footprint but also ensures that the materials are suitable for the local climate.
- ***Using Recycled or Sustainable Materials***: Where new materials are necessary, choosing recycled or sustainably sourced options can significantly reduce the environmental impact of a home. This includes materials like recycled glass or plastic, bamboo flooring, or countertops made from recycled composite materials.

Designing an energy-efficient container home involves several strategies to reduce the amount of energy needed for heating, cooling, and lighting:

- ***Insulation***: Proper insulation is crucial in a container home to prevent heat transfer. Insulating paints, spray foam, or panels made from recycled materials can be used to insulate the steel walls and roof effectively.

- ***Thermal Mass***: Incorporating materials that absorb and slowly release heat can help stabilize indoor temperatures. In container homes, this might involve designing concrete or tiled floors that absorb heat during the day and release it at night.

- ***Passive Solar Design***: Orienting the home to take advantage of natural sunlight can significantly reduce lighting and heating demands. Large, strategically placed windows can provide ample natural light and warmth during the colder months while overhangs or awnings can minimize solar gain during the summer.

Water conservation is another critical aspect of eco-friendly design, particularly in regions prone to drought. Implementing water-efficient systems in container homes can include:

- ***Rainwater Harvesting Systems***: Installing gutters and storage tanks to collect and reuse rainwater for irrigation and, if properly treated, for domestic use.

- ***Low-Flow Fixtures***: Using low-flow toilets, showerheads, and faucets to reduce water usage.

- ***Native Landscaping***: Designing landscapes with native plants and trees that require minimal irrigation and maintenance.

Maintaining a healthy indoor environment is essential for the well-being of residents. Eco-friendly design prioritizes indoor environmental quality through:

- ***Use of Non-Toxic Materials***: Selecting paints, sealants, and adhesives that contain low or no volatile organic compounds (VOCs) reduces air pollution inside the home.

- ***Natural Ventilation***: Designing the home to enhance airflow can improve air quality and reduce the need for mechanical ventilation.

Beyond the structural and material considerations, integrating green technologies like solar panels, geothermal systems, or wind turbines can further reduce a home's carbon footprint and lower utility bills.

The principles of eco-friendly design are not merely about reducing the environmental impact of a container home but enhancing its functionality, aesthetics, and livability. By adhering to these principles, designers and homeowners can create spaces that are not only efficient and sustainable but also healthful and harmonious with their natural surroundings. As we continue to face global environmental challenges, the importance of integrating these principles into every facet of home design becomes ever more apparent. Container homes, built with these ideals in mind, exemplify the potential for innovative, sustainable living solutions in the modern world.

Maximizing Space and Light

Designing a container home requires a thoughtful approach to maximize space and light, fundamental elements that transform the standard confines of a metal box into a pleasant, livable space. This challenge is crucial in container architecture, where the inherent limitations of a narrow, rectangular structure demand creativity and precision. This section delves into the strategic design choices that can enhance the spaciousness and luminosity of container homes, ensuring that these structures are not only functional and efficient but also bright and welcoming.

The typical shipping container offers a unique set of spatial constraints: generally, about eight feet in width, eight feet six inches in height, and either twenty or forty feet in length. These dimensions create a naturally elongated and narrow room, which can feel confined if not designed with care. However, these constraints also present an opportunity to innovate with space-saving solutions and light-enhancing techniques that can dramatically transform the feel of the home.

The layout of a container home is pivotal in maximizing space. Designers must consider the best ways to utilize the linear nature of containers while overcoming feelings of confinement:

- ***Open Floor Plans***: Embracing an open floor plan can make a small space feel larger and more connected. Minimizing internal walls increases the flow and versatility of the living area, making it adaptable to various activities and social settings.

- ***Multi-functional Furniture***: Incorporating furniture that can serve multiple purposes — such as Murphy beds, foldable desks, and modular sofas — can free up floor space when these elements are not in use. This adaptability is especially valuable in a container home.

- ***Custom Built-ins***: Custom cabinetry and built-ins that conform to the container's dimensions can maximize storage and maintain a sleek, uncluttered appearance. These elements can be designed to fit snugly into the home's layout, utilizing every possible inch of space.

Maximizing natural light and strategically using artificial lighting are key to transforming the interior of a container home from a cramped corridor to a bright, airy space:

- ***Large Windows and Glass Doors***: Installing large windows, glass doors, and skylights can flood the interior with natural light, making the space feel larger and more open. Placement should be considered based on the sun's path to maximize light intake during the day and to offer views that extend the eye beyond the home's physical boundaries.
- ***Reflective Surfaces and Color Palette***: Using reflective surfaces and a light color palette can enhance the sense of space. Glossy finishes on floors, ceilings, and cabinets can reflect light deeper into the home. Light wall colors, particularly whites and pastels, make the walls seem to recede, creating a feeling of openness.

- ***Smart Lighting Design***: Integrating layers of lighting ambient, task, and accent can sculpt the space visually and functionally. Recessed lighting can keep lines clean and ceilings uncluttered, while directed task lighting can highlight work areas or key features without overwhelming the space.

In container home design, the connection between indoor and outdoor spaces can significantly enhance the overall living area:

- ***Decks and Patios***: Designing decks, patios, or balconies that seamlessly connect with the interior through sliding glass doors or large openings can extend the living space outdoors, providing additional room to relax and entertain.

- ***Transitional Spaces***: Incorporating elements like covered porches or sunrooms can offer transitional spaces that blend the indoors with the outdoors, providing protected areas to enjoy natural surroundings even in less favorable weather.

The height of the container can also be utilized creatively to maximize space:

- ***Loft Areas***: Adding a loft for sleeping or storage takes advantage of the vertical space, freeing up the lower areas for living and activity spaces.

- ***High Shelving and Storage***: Installing shelving that reaches near the ceiling can draw the eye upward and make use of often under-utilized upper wall space.

Designing a container home to maximize space and light involves a multifaceted approach that incorporates thoughtful layout planning, clever use of furniture, strategic window placement, and effective lighting. These elements, when harmoniously integrated, can transform the restrictive dimensions of a shipping container into a spacious, luminous home environment. By pushing the boundaries of traditional design and embracing the challenges posed by container dimensions, architects and homeowners can create exceptionally functional, beautiful living spaces that feel open and connected to their surroundings.

Incorporating Green Technologies

The integration of green technologies into container homes is more than just a trend; it's a strategic approach to sustainable living that minimizes environmental impact while enhancing efficiency and comfort. This chapter delves into how incorporating cutting-edge green technologies can transform a shipping container home into an eco-friendly, energy-efficient dwelling. Understanding and applying these technologies is crucial for those who wish to lead a greener lifestyle and contribute positively to environmental conservation.

Green technologies in the context of container home design encompass a broad range of systems and devices designed to reduce energy consumption, minimize waste, and promote the sustainable use of resources.

These technologies are not just about harnessing renewable energy; they also involve water conservation, waste reduction, and maintaining indoor environmental quality. The goal is to create homes that are not only energy-efficient but also healthy living environments.

One of the core components of green technology in home design is the focus on energy efficiency and the utilization of renewable energy sources:

- ***Solar Power Systems***: Photovoltaic (PV) panels are among the most popular and effective technologies for container homes. These systems convert sunlight directly into electricity, which can power household appliances, lighting, and HVAC systems. The design of container homes often includes flat or slightly sloped roofs, which are ideal for solar panel installation. In addition, the portability and modular nature of containers allow for solar setups to be pre-installed off-site, reducing installation costs and complexity.

- ***Wind Turbines***: Small-scale wind turbines can be another viable option for container homes located in windy areas. These turbines generate electricity from wind energy, which can be used directly or stored in batteries for later use. When combined with solar power, wind energy can help create a truly off-grid home capable of producing its own clean energy year-round.

- ***Geothermal Heating and Cooling***: Geothermal systems use the earth's stable underground temperature to heat and cool the home, significantly reducing the need for traditional HVAC systems. While the initial installation can be costly and complex, the long-term savings and environmental benefits are substantial.

Another crucial aspect of green technology in container homes is water conservation. Innovative solutions in this area not only reduce water usage but also manage wastewater effectively:

- ***Rainwater Harvesting Systems***: These systems collect and store rainwater from roofs, which can be used for irrigation, flushing toilets, and, with proper treatment, for drinking and bathing. This technology reduces dependency on municipal water systems and lowers water bills.

- ***Greywater Recycling Systems***: Greywater refers to wastewater generated from sinks, showers, and washing machines. Recycling this water for use in flushing toilets or watering gardens can significantly reduce the water footprint of a home.

- ***Low-Flow Fixtures***: Installing low-flow toilets, showerheads, and faucets can dramatically decrease water usage without sacrificing performance. These fixtures are designed to use less water per minute, helping to conserve this vital resource.

The use of smart home technologies in container homes enhances efficiency and control over the home's environmental systems:

- ***Smart Thermostats***: These devices allow for automated temperature control and can learn the homeowner's preferences, optimizing heating and cooling schedules to save energy.

- ***Energy Management Systems***: Home energy management systems provide detailed insights into energy consumption patterns, enabling homeowners to make informed decisions about their energy use.

- ***Automated Lighting Systems***: Integrating motion sensors and automated schedules can significantly reduce energy consumption from lighting, ensuring lights are only on when needed.

In addition to specific green technologies, the choice of materials and construction methods can also contribute to a container home's sustainability:

- ***Sustainable Building Materials***: Using recycled, reclaimed, or sustainably sourced materials reduces the environmental impact of building a new home. For instance, using reclaimed wood for interior finishes or recycled steel for structural enhancements not only adds aesthetic value but also conserves resources.
- ***Insulation Materials***: High-quality, sustainable insulation materials like cellulose, wool, or polystyrene made from recycled content can improve the thermal efficiency of container homes, reducing the need for heating and cooling.

Integrating green technologies into container home design is essential for those looking to minimize their ecological footprint while maximizing comfort and efficiency. From energy production and conservation to water management and smart systems, these technologies enable homeowners to lead more sustainable lives. By thoughtfully incorporating these innovations, container homes can serve as a model for sustainable residential architecture, showcasing how modern living can be environmentally responsible and technologically advanced.

Container home

Chapter 6: Getting the Foundations Right

Types of Foundations Suitable for Container Homes

Choosing the right foundation for a shipping container home is a pivotal decision in the construction process. It influences not only the stability and durability of the structure but also affects the overall cost, construction timeline, and long-term maintenance of the home. The foundation must be specifically tailored to the unique characteristics of shipping container construction, ensuring that the structure is secure, level, and adequately insulated from moisture and other environmental factors. This comprehensive exploration of foundation types suitable for container homes will guide prospective builders through the considerations and options available, ensuring that the foundation chosen will suit their specific needs and conditions.

The need for a robust foundation in container home construction cannot be overstated. Unlike traditional buildings, container homes are constructed from steel, which, while structurally sound, can be prone to condensation and thermal bridging if not properly insulated from the ground. Moreover, the weight distribution in container homes, especially if containers are stacked or joined in an unconventional layout, requires careful consideration to prevent settling or shifting that could lead to structural issues.

Several types of foundations are commonly used in container home construction, each with its own set of advantages, challenges, and suitability depending on the project specifics such as location, soil type, and design complexity.

1. Pier Foundations

Pier foundations consist of concrete or steel piers (or columns) that are positioned at the container's corners and sometimes along its length, depending on the container configuration. This type of foundation is relatively inexpensive and quick to install. It is ideal for sites with uneven terrain or where minimal site impact is desired. Pier foundations elevate the container off the ground, preventing moisture problems and allowing for easy access to plumbing and electrical connections underneath the home.

2. Slab-on-Grade Foundations

A slab-on-grade foundation involves pouring a flat concrete pad on leveled ground. This type of foundation is one of the simplest and most cost-effective for container homes and is particularly suitable for climates where the ground does not freeze, as frost heave can be a concern. The slab provides a solid base that distributes the weight of the container evenly, reducing the risk of settling. Additionally, a slab foundation can include integral footings where the concrete is thicker under the container load points, offering extra stability.

3. Strip Foundations

Strip foundations, also known as trench foundations, involve pouring concrete into a continuous strip under all the load-bearing walls of the home. This foundation type is particularly useful for multiple container configurations where load distribution needs to be managed across several points. It is more labor-intensive than pier or slab foundations and is used in situations where additional stability is required due to soil conditions or design complexity.

4. Pile Foundations

Pile foundations are used primarily in locations with very poor soil conditions where other types of foundations might fail. They involve driving steel, concrete, or wood piles deep into the ground to reach stable soil layers. This type of foundation is the most expensive and technically demanding, typically used when building on unstable soils such as clay or sand, or in flood-prone areas.

5. Raft Foundations

Raft foundations, or mat foundations, involve a thick concrete slab that covers the entire footprint of the home, with reinforced bars for added strength. This foundation type is effective in distributing the load evenly, especially in areas with loose soil. It prevents differential settling across the structure and is ideal for larger, multi-container homes.

Selecting the right foundation for a container home involves considering several factors:

- ***Soil Type and Quality***: Soil tests can determine the bearing capacity of the soil, which influences the choice of foundation. Sandy soils might require deeper, more robust foundations like piers or piles, while clay soils benefit from wider foundations that distribute loads over a larger area.

- ***Climate***: In areas with harsh winters, frost-proof foundations are necessary to prevent the movement caused by the freeze-thaw cycle.

- ***Topography***: Sloped or uneven land might require specialized foundations like piers or a combination of foundation types to ensure level and stable construction.

- ***Budget and Longevity***: While budget constraints are significant, investing in a more robust foundation can prevent future issues and costs associated with settling or structural damage.

The foundation of a container home is as crucial as the structure itself. It ensures the home's stability, durability, and comfort. Understanding the different types of foundations available and their suitability based on the specific conditions and needs of your project is essential. By carefully planning and selecting the appropriate foundation, you can ensure that your container home remains a secure and lasting dwelling. This foundational decision not only supports the physical structure but also lays the groundwork for a home that meets all expectations of safety, comfort, and style.

DIY vs. Professional Foundation Laying

When embarking on the construction of a container home, one of the most significant decisions you will face is whether to undertake the foundation laying yourself or to engage professional services. This choice not only affects the project's budget but also its timeline, the physical safety of the structure, and the long-term stability and durability of your home. Understanding the complexities, risks, and benefits associated with both DIY and professional foundation laying is crucial for making an informed decision that aligns with your capabilities, budget, and project goals.

DIY foundation laying begins with a solid understanding of what type of foundation is best suited for your container home. This decision is influenced by various factors including soil type, climate, topography, and the overall design of the home. For instance, a simple pier foundation might be feasible for a DIY enthusiast in a temperate climate on stable, flat ground. However, more complex foundations, such as slab-on-grade or deep pile systems, often require technical expertise and specialized equipment.

One of the most compelling reasons to consider a DIY approach is cost savings. Professional foundation laying can be a significant portion of the construction budget.

By undertaking this task yourself, you can potentially save thousands of dollars. However, these savings must be weighed against the need for equipment rental, materials purchasing, and the potential cost of correcting mistakes, which can sometimes exceed the cost of hiring professionals from the start.

Laying a foundation is labor-intensive. It requires not just physical labor but also time commitment. For those who are not full-time builders, the DIY process can extend the timeline of the construction project significantly. It's important to realistically assess your own or your team's ability to dedicate the necessary time and labor to ensure the foundation is laid correctly.

If you decide to lay the foundation yourself, a substantial amount of learning and preparation is required. This might include studying building codes, learning about concrete mixing and curing, understanding grading and drainage, and knowing how to set and level forms for concrete. The internet, library books, and advice from experienced builders can be invaluable resources.

Risks and Challenges of DIY Foundation Laying

1. *Potential for Mistakes*:

The risk of errors is considerably higher with DIY foundation laying, especially for those with limited experience in construction. Mistakes in laying a foundation can lead to serious issues like uneven settling, water infiltration, and structural instability, which can compromise the safety and longevity of the home.

2. *Compliance with Building Codes*:

All construction projects, including container homes, must comply with local building codes. DIY projects are no exception. Non-compliance can lead to fines, required redoing of work, or even legal issues. It is vital to understand and adhere to these regulations.

Benefits of Professional Foundation Laying

1. *Expertise and Experience*:

Professional builders bring expertise and experience that can be critical in avoiding costly mistakes. They understand how to navigate the challenges associated with different soil types, weather conditions, and topographical features. Professionals also stay up-to-date with building codes and often have established relationships with local inspectors.

2. *Speed and Efficiency*:

Due to their skills and tools, professionals can complete foundation laying faster and with greater efficiency compared to DIY builders. This speed can be crucial if the project is on a tight schedule or if you're working within a specific time frame due to weather conditions or other external factors.

3. *Guaranteed Quality and Peace of Mind*:

Many professional services offer guarantees for their work. Hiring a reputable contractor can provide peace of mind, knowing that the foundation is secure and laid correctly, which is a solid investment for the future of your home.

Choosing between DIY and professional foundation laying for your container home involves a careful assessment of your skills, resources, and project requirements. While DIY can offer cost savings and a profound sense of achievement, the risks and potential for costly errors are significant. Professional foundation laying, though more expensive, offers expertise, efficiency, compliance assurance, and peace of mind. This decision should be based on a realistic evaluation of your abilities and the specific demands of your project, ensuring that the foundation of your container home is as strong and stable as the dreams it is built upon.

Ensuring Stability and Durability

Constructing a foundation for a container home involves much more than simply laying a solid base. It's about ensuring the long-term stability and durability of a structure that is fundamentally different from traditional buildings. This part of your project is crucial; it not only supports the weight of your home but also resists the forces of nature, from winds and earthquakes to moisture and thermal changes. Let's delve into the specifics of what makes a foundation stable and durable, focusing on the materials, techniques, and considerations that are key to building a lasting container home.

Container homes pose unique challenges and requirements for foundation stability and durability. Unlike traditional wood or brick-built homes, containers are prefabricated steel units that have specific structural and weight distribution characteristics. The foundation must accommodate these features to avoid issues such as differential settling, which can lead to structural cracks, misalignments, and other damage over time.

Several factors must be considered to ensure the foundation provides stable and durable support for a container home:

- ***Soil Analysis***: Before any foundation work begins, a thorough analysis of the soil at the construction site is essential. Soil type, density, and stability can significantly affect the choice of foundation. Sandy or clay-heavy soils may require deeper footings or special treatments to ensure stability, while rocky or compacted soils might offer a more stable base for standard foundation types.

- ***Weight Distribution***: Containers themselves are sturdy, but when modified and combined to form a larger structure, the weight distribution can change. Foundations for container homes need to accommodate these alterations, particularly if containers are stacked or joined in non-standard configurations.

- ***Climate Considerations***: Environmental factors play a significant role in foundation stability. In areas prone to freezing, foundations must extend below the frost line to prevent heaving. In regions susceptible to heavy rains or flooding, adequate drainage and waterproofing measures are critical to protect the foundation from water damage.

- ***Seismic and Wind Considerations***: In earthquake-prone areas, foundations need to be designed to withstand seismic forces. This might include the integration of flexible joints or reinforced materials. Similarly, in high wind zones, the foundation must securely anchor the containers to prevent shifting or overturning.

Enhancing the stability and durability of a container home foundation involves several techniques and best practices:

- ***Deep Foundations***: In unstable soil conditions, deep foundations such as driven piles or drilled shafts can reach stable soil layers deep underground, providing a firm anchor for the home.

- ***Reinforcement Strategies***: Adding reinforcement, such as rebar or wire mesh, within concrete foundations can significantly increase their strength and resilience to cracking under stress.

- ***Waterproofing Measures***: Applying waterproof membranes or using water-resistant materials can protect the foundation from moisture infiltration, which can weaken the structure and cause mold issues.

- ***Drainage Systems***: Proper drainage systems are crucial to direct water away from the foundation. Techniques include graded surfaces that divert water, French drains, or incorporating porous materials that facilitate water runoff.

- ***Thermal Considerations***: For regions with significant temperature fluctuations, insulating materials may be used within or around the foundation to prevent damage from thermal expansion and contraction.
Ongoing monitoring and maintenance are integral to ensuring the long-term stability and durability of a container home foundation:

- ***Regular Inspections***: Scheduled inspections can help identify potential issues before they become serious problems. Look for signs of settling, cracking, or moisture buildup that could indicate problems with the foundation.

- ***Immediate Repairs***: Addressing minor issues promptly can prevent more significant damage. Filling cracks, resealing waterproof membranes, and clearing drainage systems are all maintenance tasks that can prolong the life of a foundation.

- ***Adaptations and Upgrades***: Over time, changes in the surrounding environment or the home itself may necessitate adaptations to the foundation. Being open to making these changes can ensure the ongoing stability of the structure.

Conclusion

The foundation of a container home is the bedrock upon which the entire project rests. Ensuring its stability and durability is not just about following current standards but anticipating future challenges. By carefully planning, using appropriate materials and techniques, and committing to regular maintenance, you can ensure that your container home remains safe, secure, and beautiful for years to come. This proactive approach to foundation design and care is indispensable, safeguarding your investment and ensuring that your home continues to be a source of pride and comfort.

Chapter 7: Structural Modifications and Reinforcements

Cutting, Framing, and Joining Techniques

Constructing a container home involves significant structural modifications that transform standard shipping containers into functional and aesthetically pleasing living spaces. These modifications include cutting for windows and doors, framing to add structural support, and joining multiple containers to create a cohesive living space. This exploration of techniques ensures safety, efficiency, and durability in home construction, focusing on methods that integrate seamlessly into the broader construction process.

Before any cutting or adjustments are made, it is essential to understand the basic structure of a shipping container. Constructed primarily from Corten steel, shipping containers are designed to be incredibly robust, capable of withstanding heavy loads and harsh environments. However, modifying the container, such as cutting openings for windows or doors, can compromise its structural integrity if not executed correctly.

The process of cutting into steel containers requires precision and appropriate tools to ensure clean cuts without compromising the container's strength. Plasma cutters are typically used for their efficiency and precision in cutting steel, utilizing a high-velocity jet of ionized gas to melt the metal. For smaller openings or more accessible jobs, angle grinders and reciprocating saws can also be employed, though they may not provide the same level of precision and can require more effort.

Safety is paramount when cutting steel. Proper safety gear, including gloves, eye protection, and ear protection, is essential, and ensuring the area is well-ventilated to avoid the inhalation of metal fumes is crucial.

After cutting the necessary openings, framing is required to add support and create space for insulation and utilities. Steel framing involves using steel studs that can be either welded or bolted to the container walls, offering durability and resistance to fire and pests. Alternatively, wood framing provides ease of use and accessibility for attaching interior finishes and can be affixed using welded brackets or self-tapping screws.

Joining multiple containers is a critical step in container home construction, requiring careful alignment and secure fastening to ensure the structural integrity of the home. Welding offers a permanent solution by bonding the steel of one container to another, creating a robust continuous structure. For less permanent solutions or when future disassembly might be required, bolting provides adequate security and allows for easier modification later. Clamping is another method used primarily for temporary setups or when containers need to be moved frequently.

Maintaining the structural integrity of the container during modifications is paramount. It is crucial to understand which parts of the container are load-bearing and to avoid weakening these areas during modifications. Areas that have been cut or altered should be reinforced to handle the stresses the container will face as part of a home. Moreover, all modifications should comply with local building codes and standards, which may specify requirements for maintaining structural integrity.

The process of cutting, framing, and joining shipping containers for home construction requires not only technical skill but a deep understanding of the materials and methods involved. By carefully planning and executing these modifications, builders can ensure that the container home remains safe, stable, and durable. This approach to container modification not only preserves the structural integrity of the original materials but also adapts them to new uses, providing sustainable and innovative housing solutions. Through meticulous attention to detail and adherence to safety and building standards, a container home can be transformed from a simple steel box into a secure, efficient, and welcoming living space.

Maintaining Structural Integrity

Transforming shipping containers into habitable spaces necessitates significant structural modifications. Ensuring the ongoing structural integrity of these modified containers is paramount to the safety, longevity, and functionality of the resulting home. This exploration delves into the principles and practices crucial to maintaining structural integrity through various stages of container home modifications. Key focus areas include assessing and reinforcing structural components, considering the impact of modifications on the container's original design, and adhering to building codes and standards.

Structural integrity refers to the ability of a building to withstand the intended loads and forces over its lifespan without experiencing failure or excessive deformation. For container homes, this involves retaining the strength and rigidity of the containers even after they have been cut, joined, and transformed into living spaces. This is critical because any compromise in the container’s structure can lead to problems such as leaks, misalignments, or even catastrophic structural failures.

Container modifications often involve cutting large openings for doors, windows, and room transitions, which inherently weaken the structural framework of a steel box. Before making any cuts, it’s essential to understand how these modifications will affect the overall load-bearing capacity of the container. Areas that will experience increased stress or load redistribution should be identified. Plans must be made to reinforce these areas to handle the altered load paths after determining where the container will be cut.

Various techniques can be used to reinforce a container, depending on the extent and nature of the modifications. The most common method of reinforcement is the addition of steel beams or plates at cut sections to help redistribute loads and provide support where the original strength has been compromised. In some cases, especially in non-load-bearing modifications, structural adhesives are used to increase the rigidity and strength of joined sections. These adhesives are designed to withstand high loads and can be used in conjunction with mechanical fasteners.

For extensive modifications, welding additional steel frameworks inside the container can provide the necessary support. This is often required when containers are stacked or when creating large open spaces that remove much of the container's inherent structural support.

Every modification made to a container affects its structural integrity in some way. Proper analysis must be conducted to ensure that any cuts made do not compromise the container's ability to support itself and any additional loads from roofs, snow, or even other stacked containers. When containers are joined to create larger spaces, the points of connection become critical stress points. Ensuring these joints are as strong as, or stronger than, the original structure is vital. Modifications can also affect the long-term durability of the structure, particularly in terms of corrosion and fatigue. Protective coatings and regular maintenance can help mitigate these issues.

Compliance with local building codes and standards is not just a legal requirement but also a guideline for maintaining safety through structural integrity. These codes provide detailed requirements for all aspects of building construction, including minimum structural requirements set to ensure that homes can withstand typical loads and forces expected in their geographic location. Building codes generally include safety margins that ensure structures can handle greater loads than those they are expected to encounter during their use. Adhering to these standards often requires periodic inspections by certified professionals, ensuring ongoing compliance and safety.

Maintaining structural integrity during the modifications of shipping containers for home construction requires a comprehensive understanding of structural engineering principles, meticulous planning, and precise execution. By carefully assessing the impacts of modifications, reinforcing the structure where necessary, and adhering to building codes, builders can ensure that container homes are safe, durable, and capable of standing the test of time. These practices not only protect the inhabitants but also enhance the value and longevity of the home, ensuring that it remains a sustainable and sturdy dwelling for years to come.

Waterproofing and Rust Prevention

Building a home from shipping containers involves addressing the unique challenges posed by the material itself—steel. While steel offers robust structural benefits, it is also prone to issues like corrosion and water ingress, which can significantly undermine the durability and safety of the home. Effective waterproofing and rust prevention are critical to ensuring that these potential vulnerabilities are addressed, thus extending the lifespan of a container home and maintaining its structural integrity. This comprehensive guide explores the methods and practices crucial to safeguarding container homes against water damage and corrosion.

Shipping containers are designed to withstand the harsh conditions of ocean travel, which exposes them to saltwater, intense winds, and fluctuating temperatures. However, when repurposed as building blocks for homes, these containers can be susceptible to different environmental stressors. Without proper treatment, the steel can rust, and joints or modifications may allow water to enter, leading to issues like mold, structural weakening, and insulation deterioration.

Waterproofing container homes involves key strategies that protect against water penetration, ensuring the home remains dry and free from water-induced damage. One of the most straightforward and essential steps in waterproofing is the application of high-quality sealants. Sealants should be applied to joints, seams, and around openings such as doors and windows where water might infiltrate. Silicone or polyurethane sealants are commonly used because of their flexibility, durability, and strong adherence to metal.

The roof of a container home often requires additional waterproofing measures due to its exposure to rain. Applying a waterproof roofing membrane or coating can prevent water accumulation and leakage. These membranes not only add a waterproof layer but also can reflect sunlight, helping to reduce heat absorption.

Specialized protective paints and coatings that provide a waterproof barrier over the steel surfaces are also used. These products often include rust inhibitors that add an extra layer of protection against corrosion. Epoxy and ceramic coatings are popular options that provide both waterproofing and aesthetic enhancement.

Ensuring good drainage around the container home is vital. This might involve the installation of gutters and downspouts to channel water away from the home, as well as grading the land around the foundation to prevent water pooling.

Preventing rust is critical to maintaining the structural strength and aesthetic appearance of a container home. Proper preparation of the container's surface is essential before applying any paints or coatings. This typically involves removing any existing rust using sandblasting or mechanical brushing, followed by cleaning the surface to remove oils and contaminants.

Once the surface is prepared, applying a rust-inhibitive primer is crucial. This primer acts as a protective layer that prevents oxidation of the steel, which is the chemical process that leads to rust.

Regularly inspecting the container for signs of rust or breaches in the waterproofing layers is important. Early detection and repair of small issues can prevent them from developing into major problems.

To ensure effectiveness, waterproofing and rust prevention should be integrated into the construction process from the start. During the design phase, consider factors such as climate, site exposure, and local weather patterns. Design elements like overhangs or covered porches can provide natural protection against the elements.

Choose materials that complement the protective strategies, such as using stainless steel fasteners or hardware that are less likely to corrode.

During construction, maintain high-quality standards for any modifications or installations that penetrate the container's surface. Ensure that all cuts are properly sealed and treated to prevent rust.

Waterproofing and rust prevention are not merely about applying products; they require a systematic approach that starts from the initial design and continues through to regular home maintenance. By implementing thorough waterproofing measures and proactive rust prevention strategies, the longevity and durability of a container home can be significantly enhanced. This not only protects the investment but also ensures that the home remains safe, comfortable, and aesthetically pleasing for years to come.

Such meticulous attention to detail in protecting against water and corrosion is essential for anyone looking to transform a shipping container into a lasting and sustainable home.

Chapter 8: Insulation and Climate Control

Comparing Insulation Materials for Efficiency

Insulation is a fundamental component in the construction of container homes, pivotal for creating a comfortable living environment regardless of external climate conditions. This chapter delves into the various insulation materials available, comparing their efficiency and suitability for use in container homes. It aims to provide an in-depth understanding of how different insulation options can affect energy consumption, indoor air quality, and overall comfort

Insulation in a container home serves multiple critical functions. It acts as a barrier against heat loss in colder climates and heat gain in warmer climates, thereby significantly reducing the need for active heating and cooling systems. Proper insulation not only improves comfort but also enhances energy efficiency, leading to lower utility bills and a reduced environmental impact. Additionally, insulation contributes to sound dampening, an important consideration given the inherent acoustic properties of steel containers.

A variety of insulation materials are available, each with its own set of properties that can be beneficial or detrimental depending on the specific needs and environmental conditions of the container home. Here we explore some of the most commonly used insulation types:

- ***Fiberglass Insulation***: One of the most traditional forms of insulation, fiberglass is made from fine glass fibers and is commonly used in batts, rolls, or loose-fill. It is relatively inexpensive and has a good thermal resistance value per inch. However, fiberglass can be harmful to install without proper safety gear as the tiny glass particles can irritate the skin, eyes, and respiratory system.

- ***Spray Foam Insulation***: Spray foam is an effective option for container homes because it provides both a high R-value per inch and a seamless moisture barrier. It can be sprayed directly onto the interior or exterior walls of a container, expanding to fill all gaps and crevices. This type of insulation is excellent for preventing thermal bridging—a common issue where heat bypasses the insulated areas of a structure via materials with higher thermal conductivity.

- ***Rigid Foam Boards***: Rigid foam offers a high insulation value and moisture resistance. These boards can be cut to fit precisely between the framing elements of a container home, making them a popular choice for floors, walls, and roofs. Rigid foam is more expensive than fiberglass but typically offers superior thermal resistance and durability.

- ***Reflective Insulation***: This material is designed to reflect radiant heat rather than absorb it, making it particularly useful in hot climates. Reflective insulation is often used in conjunction with other insulation types to increase the overall energy efficiency of a home.

- ***Recycled Cotton or Denim***: For those seeking an eco-friendly option, recycled cotton or denim provides a sustainable choice. These materials are typically treated with boric acid to make them fire and pest-resistant. While offering moderate insulation properties, they are also safer to handle and install than fiberglass.

When choosing insulation for a container home, several performance factors must be considered:

- ***Thermal Resistance (R-value)***: This measures how well a material can resist heat flow. The higher the R-value, the better the insulation performance. It is crucial to choose an insulation with the appropriate R-value for your climate and the specific placement within your home.

- ***Air and Vapor Barriers***: Some insulation materials also act as air or vapor barriers, which can be crucial in preventing moisture-related issues such as mold and mildew. In climates with high humidity, selecting an insulation material that offers these additional properties can be beneficial.

- ***Environmental Impact***: For many homeowners, the environmental impact of insulation materials is a significant consideration. Materials like recycled cotton, wool, and certain types of foam with low volatile organic compound (VOC) emissions can offer sustainable benefits without compromising on efficiency.

- ***Cost-Effectiveness***: While the initial cost of the insulation material is an important factor, the long-term savings through reduced energy bills should also be considered. Investing in a more expensive insulation that offers higher energy efficiency might result in greater overall savings.

Selecting the right insulation material for a container home involves a careful evaluation of various factors, including thermal performance, moisture resistance, environmental impact, and cost-effectiveness. By understanding the unique properties and benefits of each insulation type, builders and homeowners can make informed decisions that enhance the comfort, sustainability, and efficiency of their container homes. Effective insulation is key to transforming a steel container into a warm and inviting living space, and with the right materials, it can become a cozy retreat that stands the test of time.

Innovative Solutions for Heating and Cooling

Creating a comfortable living environment in a container home requires more than just well-placed insulation; it demands a thoughtful approach to heating and cooling that addresses the unique challenges of these steel structures. Traditional HVAC systems might not always be suitable or efficient for the compact spaces of container homes. This part of the chapter explores innovative solutions for climate control in container homes, focusing on sustainable, efficient, and effective methods to maintain optimal indoor temperatures throughout the year.

Due to their metallic construction, shipping containers are naturally prone to extreme temperature fluctuations. In the absence of effective climate control solutions, these homes can become uncomfortably hot in summer and bitterly cold in winter. The challenge is to implement heating and cooling solutions that not only address these fluctuations but do so in an energy-efficient manner that is also compatible with the spatial and aesthetic constraints of container homes.

Heat Pumps: A Versatile Solution

One of the most effective solutions for both heating and cooling in container homes is the use of heat pumps. These devices operate by extracting heat from outside air, even in cold weather, to heat the home, and can reverse the process in the summer to cool the interior. Heat pumps are highly efficient because they transfer heat rather than generate it by burning fuel. Mini-split systems, in particular, are well-suited to container homes because they do not require ductwork and can be installed with minimal intrusion into the structural integrity of the container.

Solar Air Conditioning

Harnessing solar energy for cooling can significantly reduce the energy costs associated with traditional air conditioning systems. Solar air conditioning comes in two main forms: solar thermal systems, which use heat from the sun to drive a cooling process, and photovoltaic systems, which use solar panels to generate electricity to power an air conditioner.

Both systems can be highly effective in container homes, particularly in sunny climates, and can be integrated into the design of the home to maintain aesthetic coherence.

Evaporative Coolers

In areas with a dry climate, evaporative coolers offer an energy-efficient alternative to traditional air conditioning. Also known as swamp coolers, these devices cool outdoor air by passing it over water-saturated pads, causing the water to evaporate and reduce the air temperature. For container homes, portable or window-mounted evaporative coolers can provide significant cooling during hot months, using much less electricity than standard air conditioners.

Geothermal Systems

Although more expensive and complex to install, geothermal heating and cooling systems offer a highly sustainable option for container homes. These systems use the earth's stable underground temperature to regulate the temperature of the home, providing both heating and cooling. The installation involves placing a series of pipes, known as a ground loop, beneath the ground. Fluid circulating through these pipes absorbs earth's heat in the winter and dispels heat in the summer, keeping the home comfortable year-round.

Radiant Floor Heating

Radiant floor heating is another effective method for heating container homes. This system involves installing pipes or electric heating elements beneath the floor surface, which evenly distribute heat throughout the space. The warmth from the floor rises, heating the entire room efficiently. Radiant heating is particularly effective in container homes, as it eliminates the issues of cold floors, providing a cozy living environment and reducing the circulation of dust and allergens associated with forced-air systems.

Smart Thermostats

Integrating smart thermostats into the climate control systems of container homes can significantly enhance energy efficiency. These devices allow homeowners to automate heating and cooling schedules based on their daily routines and adjust the temperature remotely via smartphones. Smart thermostats can learn the homeowners' preferences over time and adjust settings to maximize comfort and efficiency.

Often, the best approach to climate control in container homes involves combining several of these systems. For instance, using a heat pump in conjunction with solar panels can cover both heating and cooling needs while minimizing the environmental impact. Adding a smart thermostat can further enhance these systems' efficiency by optimizing their operation based on real-time weather conditions and the homeowners' presence.

Choosing the right heating and cooling solutions for a container home requires careful consideration of the climate, the home's layout, and the residents' lifestyle. By selecting innovative technologies that align with these factors, homeowners can enjoy a comfortable, sustainable living environment. These solutions not only address the thermal challenges posed by the metal structure but also offer efficient and effective climate control, making container homes a viable option for eco-conscious living in a variety of climates.

Managing Humidity and Preventing Condensation

Container homes, with their unique structural properties and compact design, present specific challenges when it comes to managing indoor humidity and preventing condensation. Excess moisture in a container home can lead to a host of problems, including mold growth, structural deterioration, and a decrease in indoor air quality. To ensure a healthy and durable living environment, it's crucial to implement a comprehensive approach that combines proper insulation, adequate ventilation, and moisture control strategies.

Understanding the sources of humidity in container homes is the first step toward effective moisture management. Humidity can come from external sources such as ambient air, especially in regions with high moisture levels, or from internal activities like cooking, bathing, and drying clothes. Effective management of humidity involves identifying these sources and taking steps to mitigate their impact.

Proper insulation plays a critical role in managing humidity levels. It helps maintain a consistent internal temperature, reducing the likelihood of condensation, which occurs when warm, moist air comes into contact with a cold surface. Using a vapor barrier in conjunction with insulation is essential. A vapor barrier resists the passage of moisture through wall assemblies and prevents moisture from condensing within the insulation layer, which can compromise its effectiveness and lead to mold growth.

Ventilation is key to controlling humidity and preventing condensation. There are several ventilation strategies that are effective in container homes:

Natural ventilation uses the strategic placement of windows, vents, and openings to facilitate the flow of air through the home. This method is energy-efficient and works well in climates where the outdoor air is not excessively humid.

Mechanical ventilation systems, such as exhaust fans and heat recovery ventilators, are necessary in regions with high humidity or for homes with limited opportunities for natural ventilation. These systems actively remove moist air from high humidity areas like kitchens and bathrooms.

In areas with extreme humidity, or during certain times of the year when humidity levels rise, using a dehumidifier can help maintain comfortable humidity levels inside the home. Dehumidifiers can be portable for use in specific areas, or whole-home systems can be integrated into the HVAC system.

Selecting moisture-resistant materials for construction and interior finishes can further aid in managing humidity. Materials such as ceramic tiles, water-resistant paint, and composite or vinyl flooring are excellent choices for moisture-prone areas like bathrooms and kitchens. These materials prevent the absorption of moisture into surfaces, reducing the potential for mold growth and material degradation.

Regular maintenance and monitoring are crucial to effectively managing humidity and preventing condensation in a container home. Homeowners should regularly inspect their home for signs of condensation, mold, or moisture damage, especially in hidden areas like behind furniture or in corners. Ensuring that all ventilation systems are clean and functioning correctly is also important; filters should be replaced and equipment serviced as recommended by manufacturers to maintain optimal performance.

Using hygrometers to monitor indoor humidity levels can provide valuable information for adjusting lifestyle habits and ventilation settings to maintain optimal humidity levels. In conclusion, effectively managing humidity and preventing condensation in container homes requires a multifaceted approach. By carefully planning and implementing moisture management strategies such as proper insulation, robust ventilation, and the use of moisture-resistant materials, homeowners can protect their investment and ensure that their container home remains a comfortable, healthy place to live.

This proactive approach not only preserves the structural integrity of the home but also enhances the overall quality of life for its inhabitants.

Chapter 9: Interior Design and Optimization

Space-Saving Ideas and Storage Solutions

Designing the interior of a container home presents unique challenges and opportunities, particularly in maximizing space and integrating effective storage solutions. The compact and linear nature of shipping containers demands a creative approach to interior design that includes foldable, transformable, and dual-purpose elements to enhance functionality and foster a sense of spaciousness.

Container homes typically offer a limited footprint with dimensions ranging from twenty to forty feet in length and about eight feet in width. This restricted space poses significant design challenges that can be effectively addressed by utilizing multi-functional furniture and an innovative layout that maximizes the perception of space. These homes require a thoughtful approach to interior design, where every piece of furniture and storage solution plays multiple roles.

Enhancing the feeling of spaciousness in a container home often involves utilizing the vertical space. Installing shelves and cabinets near the ceiling frees up valuable floor space while providing ample storage. This strategy also draws the eye upward, creating a sense of height and volume in the room. Loft beds are particularly effective in smaller container homes. By elevating the sleeping area, the space beneath can be used for living, dining, or additional storage, effectively doubling the utility of a single area. Wall-mounted desks and tables that fold away when not in use keep the living area flexible and adaptable to different needs.

Effective storage solutions are vital in container homes to prevent clutter and maintain an organized environment. Under-floor storage is an ingenious solution that uses the space beneath the floor for hidden drawers or trap doors, providing additional storage without impacting the visible living space.

Built-in furniture with storage, such as sofas with storage compartments underneath or beds with built-in drawers, serves dual purposes and saves space. Magnetic walls or modular pegboard systems for hanging tools, kitchen utensils, or office supplies organize items without taking up valuable counter or drawer space.

The unique structure of container homes often results in unconventional nooks and crannies that can be cleverly transformed into useful spaces. Custom shelving or cabinets fitted in corners utilize these often-underused areas effectively. Narrow spaces between appliances or fixtures can be fitted with pull-out storage for spices, pantry items, or cleaning supplies, making the most of every available inch. Additionally, if the container home design includes large windows, the space beneath the window can be turned into a bench with storage underneath. This not only provides a cozy reading nook but also adds to the home's storage capacity.

The interior design of a container home requires balancing aesthetics with functionality. Implementing space-saving furniture and innovative storage solutions allows these homes to become exemplars of efficient and stylish living spaces. Embracing the unique dimensions and characteristics of container homes and viewing each square foot as an opportunity to enhance livability and functionality is key. With careful planning and creative thinking, the interior of a container home can be transformed into a versatile, organized, and welcoming space that maximizes both comfort and style.

Selecting Materials and Finishes

When designing the interior of a container home, selecting the right materials and finishes is crucial not only for achieving a desired aesthetic but also for ensuring durability and comfort. The choices made in this regard can profoundly affect the daily living environment, influencing everything from air quality to maintenance levels. This detailed discussion will explore the considerations involved in choosing materials and finishes for container homes, aiming to provide insights that balance style, sustainability, function, and budget.

The materials used in the interior design of a container home play a significant role in creating an inviting and practical space. Beyond aesthetic appeal, materials must be chosen with consideration for their environmental impact, durability, and interaction with the unique metal structure of a container. For example, the right choices can help mitigate issues commonly associated with metal homes, such as thermal conductivity and noise levels.
Several factors influence the selection of materials and finishes in container homes:

- ***Durability and Maintenance***: Materials should be durable enough to withstand the wear and tear of everyday use while requiring minimal maintenance. This is particularly important in high-traffic areas or surfaces prone to frequent contact.

- ***Climate Adaptability***: Materials should be appropriate for the home's climate, contributing to the insulation and overall comfort. For instance, in humid climates, mold-resistant materials like treated wood or certain types of metal and synthetic finishes can be crucial.

- ***Sustainability***: For many homeowners, using environmentally friendly materials is a priority. Recycled, reclaimed, or sustainably sourced materials are not only eco-conscious choices but can also add character and uniqueness to a home.

- ***Aesthetic Compatibility***: The materials and finishes need to complement the overall design theme of the home, whether it's modern, industrial, rustic, or another style. This coherence helps create a visually appealing and harmonious space.
Choosing materials for container homes involves a nuanced understanding of their properties and best use cases:

- ***Wood***: Wood is a popular choice for flooring, walls, and ceilings because of its warmth and natural aesthetic. It can be used in its raw form for a rustic look or treated and finished for a more refined appearance. In container homes, wood can also help to soften the industrial feel of the metal.

- ***Metal***: Additional metal elements can be incorporated to complement the container's original material, maintaining an industrial aesthetic while adding functionality. Stainless steel, for example, is excellent for kitchens due to its resistance to rust and easy cleaning.

- ***Concrete***: Polished concrete flooring is a durable and stylish option that pairs well with the industrial theme of container homes. It is also excellent for thermal mass in passive solar home designs.

- ***Glass***: Using glass for interior partitions or additional windows can enhance natural lighting, helping to make small spaces appear larger and more open. Frosted or textured glass can also be used for privacy and style.

- ***Textiles and Fabrics***: Soft furnishings are essential for comfort and warmth. Textiles can be used effectively for window treatments, cushion covers, and area rugs, providing opportunities to introduce color, texture, and pattern into the space.
The finishes applied to these materials also play a critical role in the functionality and aesthetics of the interior:

- ***Paints and Coatings***: Non-toxic paints and coatings are preferable, especially in the confined spaces of a container home. Low-VOC (volatile organic compounds) products help maintain indoor air quality.

- ***Stains and Sealants***: Wood used inside container homes should be treated with stains and sealants that protect it from moisture and wear while enhancing its natural beauty.

- ***Polishes and Waxes***: These can be used to protect and enhance the natural features of materials like concrete and metal, adding a layer of shine and resistance to physical and environmental damage.

Selecting the right materials and finishes for a container home is a delicate balance of practical considerations and personal style. By carefully considering durability, climate adaptability, sustainability, and aesthetic compatibility, homeowners can create a living space that is both functional and visually pleasing. Thoughtful material selection not only enhances the living environment but also ensures that the home is sustainable and easy to maintain, making it a comfortable and stylish sanctuary.

DIY Decor and Furnishing Your Container Home

Decorating and furnishing a container home with a DIY approach allows homeowners to infuse their spaces with personal style and creativity. This method is not just about customization; it's about harnessing innovation to create a home that is both stylish and functional. DIY projects in container homes can include everything from furniture construction to inventive decor enhancements that make the most of the home's compact layout.

The essence of DIY in home decor extends beyond simple personalization. For container homeowners, it involves adapting and repurposing materials, crafting custom solutions, and applying inventive decorating techniques that maximize space and reflect personal taste. This approach is particularly useful in container homes where traditional decorating solutions may not always be feasible or effective.

Selecting and building furniture that complements the style and scale of a container home is crucial. Homeowners often opt for compact, multi-functional furniture that serves dual purposes—such as ottomans with internal storage or sofa beds that transform living areas into guest accommodations. Custom-built furniture is particularly advantageous as it allows for the creation of pieces that fit perfectly within the unique dimensions of container spaces. For example, constructing a Murphy bed can transform a bedroom into a multifunctional space, providing room for other activities when the bed is stowed away.

In DIY projects, the choice of materials can significantly influence both the aesthetics and functionality of a container home's interior. Reclaimed pallet wood, for instance, can be transformed into flooring, wall paneling, or furniture, offering a rustic charm and sustainability. Industrial pipes can be repurposed into stylish structural elements like curtain rods or shelving supports, enhancing the industrial chic of the container home. Fabrics and textiles are also pivotal in adding color, texture, and warmth, with homemade curtains, throws, or cushion covers introducing vibrancy and comfort.

The strategic use of color and lighting can profoundly transform the interior environment of a container home. Opting for light colors on walls can help reflect natural light, making spaces feel larger and more open. DIY lighting solutions, such as under-cabinet LED strips or custom-made lampshades, can provide essential lighting while adding a personal touch.

DIY art projects allow homeowners to express their style uniquely and can be adapted over time as preferences evolve. Creating artwork or wall hangings can provide focal points in each room and enhance the overall aesthetic. Upcycling old items into decor, such as converting an old ladder into a bookshelf or repurposing jars as light fixtures, adds both a creative and eco-friendly element to the home.

Practical DIY projects are particularly beneficial in container homes, addressing specific challenges such as space limitations and insulation. Handmade window treatments can control light and improve thermal efficiency, while DIY space dividers like bookcases or fabric screens can help define different areas in an open floor plan without structural changes.

Conclusion

DIY decor and furnishing enable container homeowners to create spaces that reflect their individual style and meet their functional needs. Through thoughtful selection and customization of furniture, innovative use of materials, strategic choices in color and lighting, and personalized art and decorations, container homes can be transformed into beautifully styled, functional living spaces.

This hands-on approach not only saves costs but also strengthens the connection homeowners have with their living environment, making each container home a unique and intimate expression of its inhabitants. This personalized touch brings life to the compact dimensions of a container home, turning a simple steel box into a cozy and inviting space.

3D rendering of a beautiful modern container house with a pool and night lighting

Chapter 10: Plumbing, Electrical, and Utilities

Basic Plumbing and Electrical Setup for Beginners

Transforming a shipping container into a livable home requires setting up basic plumbing and electrical systems, a task that might seem daunting to beginners. However, with a clear understanding and the right approach, even novices can successfully install the essential infrastructure needed to turn a bare container into a comfortable and functional living space. This detailed guide provides insights into the fundamental principles of plumbing and electrical setups specifically tailored for container homes, emphasizing safety and simplicity.

Plumbing in a container home involves creating systems to bring in clean water and remove waste water. Since container homes often have limited space and fewer access points compared to traditional homes, precise planning is crucial. The layout should start with deciding the locations of the kitchen, bathroom, and other areas that require water, ideally aligning them closely to minimize the length of piping, which can save on materials and simplify the system.

The installation process includes setting up water inlets where the main water line enters the home. This setup is crucial for distributing water to various fixtures like faucets and toilets. Additionally, waste water outlets or drains must be carefully positioned to ensure they comply with local building codes, which often dictate specific requirements for pipe slopes to facilitate efficient wastewater removal.

Choosing a water heating solution is another significant aspect of plumbing. In the constrained spaces of a container home, compact on-demand water heaters are popular because they provide immediate hot water, save energy, and conserve space.

Electrical installations in a container home include wiring for lighting, outlets, and appliances. Mapping out where all electrical devices and fixtures will be placed helps in planning the necessary circuits and their locations. It's wise to separate lighting and power circuits and to have dedicated circuits for high-power appliances.

Running electrical wires typically involves installing conduits to protect the wires from the metal edges of the container and to prevent electrical hazards. Wires should extend from the main breaker box to each outlet, switch, and appliance through these conduits.
Placing outlets and switches requires strategic consideration to ensure they meet the functional needs of the home while maintaining aesthetic quality. Outlets should be positioned to accommodate all appliances and general utility needs, and switches should be easily accessible, preferably near room entrances.

Safety Considerations

Handling plumbing and electrical systems safely is paramount, especially for beginners:
- Using proper tools and materials that meet safety standards is essential, particularly for electrical work where insulated tools and wearing rubber-soled shoes can prevent accidents.
- A clear understanding of local building codes and regulations is crucial. These codes are in place to ensure that building practices are safe and effective.
- After installation, having plumbing and electrical systems inspected by a professional ensures everything is up to code and safe for use.

Installing basic plumbing and electrical systems in a container home is achievable for beginners with careful planning and a focus on safety. Following a structured approach helps simplify what can initially seem like a complex task. Adhering to safety practices and local codes not only ensures the functionality of these systems but also secures the comfort and sustainability of the container home. By empowering homeowners with the knowledge to perform these installations, the process transforms a mere steel container into a well-equipped living space.

Solar Power and Alternative Energy Sources

Embracing solar power and other alternative energy sources is becoming increasingly popular among container home owners, who see these options not only as environmentally conscious choices but also as practical and cost-effective solutions for energy management. This discussion explores the integration of solar power and other renewable energy sources into container homes, detailing how these systems enhance sustainability and energy independence.

Container homes, with their compact and efficient design, are ideally suited for solar panel installations. The flat and unobstructed roofs typical of shipping containers provide perfect platforms for solar panels, allowing homeowners to maximize sunlight exposure. Solar power systems in these homes typically include photovoltaic (PV) panels, an inverter, battery storage, and a charge controller. The PV panels convert sunlight into electricity, which is then converted from direct current (DC) to alternating current (AC) suitable for home use. Batteries store excess electricity for later use, ensuring power availability during nighttime or cloudy days, while a charge controller manages the flow of energy to the batteries, enhancing system efficiency and battery longevity.

When installing a solar power system, orientation and tilt of the solar panels are critical for optimal sun exposure. In the northern hemisphere, panels should face south and be tilted in accordance with the latitude to maximize energy capture throughout the year. While the initial cost of solar power installation can be high, the potential long-term energy cost savings and available tax incentives often make solar power a financially viable option.

Beyond solar power, several other renewable energy options can be utilized in container homes. Small-scale wind turbines can complement solar power systems, especially in areas with consistent wind patterns. These turbines can help offset energy needs that solar power alone may not fully meet, particularly in cloudy or less sunny conditions. For homes located near a water source, micro-hydro power systems can provide a consistent and reliable energy supply. These systems generate power by harnessing the kinetic energy of flowing water, making them a powerful option for off-grid living.

Geothermal heating and cooling systems leverage the earth's stable underground temperature to regulate a home's climate. Although installation can be complex and the upfront costs are significant, the long-term benefits include substantial energy savings and minimal environmental impact.

Many container home owners opt for hybrid energy systems that combine solar, wind, and micro-hydro power to achieve maximum energy independence and efficiency. These systems ensure a consistent energy supply and reduce dependency on any single source, providing robust solutions for off-grid living.

Advanced energy management systems can significantly enhance the use of these energy sources by monitoring and controlling energy usage based on availability and demand. Such systems adjust energy consumption automatically, ensuring optimal efficiency and preventing energy wastage.

Despite the focus on renewable resources, having a backup generator remains a practical consideration, especially in remote or off-grid settings. A generator can provide emergency power during extended periods of insufficient renewable energy supply, ensuring continuous power availability.

The adoption of solar power and other alternative energy sources in container homes is a testament to the evolving nature of residential construction towards more sustainable practices. These energy solutions not only support environmental sustainability but also offer practical benefits in terms of reducing energy costs and promoting independence from traditional power grids. As renewable energy technology continues to advance, container homeowners are well-positioned to benefit from these innovations, showcasing how modern living spaces can integrate seamlessly with eco-friendly energy solutions. By investing in the right technologies and committing to sustainable practices, container homes can achieve remarkable energy efficiency and a significantly reduced ecological footprint, setting a positive standard for future residential developments.

Water Collection and Waste Management Systems

Incorporating efficient water collection and waste management systems is essential for sustainable living, particularly in the context of container homes. These systems not only serve functional purposes but also significantly enhance the environmental footprint and autonomy of the residence. This detailed discussion delves into the innovative methods suitable for container homes, focusing on rainwater harvesting for water collection and the utilization of composting toilets and greywater systems for effective waste management.

Rainwater harvesting is a standout option for container homes, especially those situated in remote or off-grid locations. This system involves capturing rainwater from the container roof, which is then funneled into a storage tank through a network of gutters and downspouts. Filters are employed to remove debris and contaminants before the water reaches the tank, ensuring it is clean for household use. The capacity of the storage tank is dependent on factors such as the area of the roof, local rainfall patterns, and the water needs of the household.

Once collected, rainwater can be purified and used for drinking, or it can serve other household needs such as irrigation, flushing toilets, and washing clothes, depending on the level of purification it undergoes. Simple purification systems might include sediment filters, carbon filters, and UV light treatment, each playing a role in ensuring the water's safety and usability. It's important for homeowners to consider local regulations concerning rainwater harvesting, as some regions have specific laws and guidelines governing its use. Environmentally, capturing rainwater reduces reliance on municipal water systems and minimizes stormwater runoff, which can decrease erosion and water pollution.

The waste management strategy in a container home must effectively handle both greywater and blackwater to ensure environmental responsibility and practicality. Greywater, which originates from sinks, showers, and washing machines, can be treated on-site and reused for non-potable purposes such as garden irrigation and flushing toilets. This usually involves separating greywater from blackwater, which is waste from toilets, and filtering it to remove impurities.

For those in remote or off-grid container homes, composting toilets are an excellent alternative to traditional sewage systems. These facilities treat human waste through natural processes that decompose the waste into compostable material, suitable for use as fertilizer. This approach not only significantly reduces water usage but also eliminates the need for a septic system.

Solid waste management in container homes involves reducing waste output and setting up effective recycling and composting programs. Homeowners can manage organic waste by setting up a system to compost kitchen scraps and other biodegradable materials, which can then be used to enrich the soil in gardens, further promoting sustainability.

By integrating water collection with waste management systems, homeowners can create a highly sustainable and efficient living environment. Using greywater to irrigate a garden not only recycles the water but also supports home food production, adding a layer of self-sufficiency. Similarly, combining rainwater harvesting with composting toilets can maximize water efficiency and minimize waste, creating a cycle of sustainability within the home.

Maintenance and monitoring of these systems are critical to their success. Homeowners should regularly check the levels of collected rainwater, the effectiveness of filtration systems, and the status of the composting process to ensure everything functions optimally. Engaging with the local community and adhering to regulatory requirements can also help in the smooth implementation and operation of these systems.

Incorporating advanced water collection and waste management systems transforms container homes into exemplars of sustainable living. These systems not only meet the functional needs of the home but also significantly reduce its ecological impact. By carefully planning and implementing rainwater harvesting and comprehensive waste management strategies, container homeowners can enjoy a home that is not only efficient and self-sufficient but also environmentally responsible, setting a positive example for sustainable residential design.

Chapter 11: Exterior Design and Landscaping

Enhancing Curb Appeal with Exterior Finishes

The exterior design of a container home is not only about aesthetics but also about integrating the structure into its surroundings while enhancing its curb appeal. This chapter delves into the various strategies and materials that can elevate the external appearance of container homes, transforming them from industrial boxes into captivating, modern residences. The focus will be on selecting exterior finishes that complement the landscape and reflect the homeowner's style, while also ensuring durability and sustainability.

Curb appeal refers to the attractiveness of a property as viewed from the street, and it plays a crucial role in shaping first impressions. For container homes, which often start as unassuming steel boxes, the challenge is to creatively use exterior finishes to craft an appealing and inviting home facade. Effective curb appeal not only enhances the aesthetic value of the home but can also significantly increase its market value and appeal to potential buyers.

The choice of materials for the exterior of a container home is critical and should reflect the home's architectural style, the climate, and the homeowner's personal taste. Here are some popular materials that are commonly used for enhancing the exterior of container homes:

- ***Wood Cladding***: Wood offers warmth and natural beauty, making it a popular choice for softening the industrial look of shipping containers. It can be used in its natural form to add rustic charm or painted in bold colors to create a more contemporary appearance. When selecting wood, it's important to consider its durability and maintenance needs, especially in harsh weather conditions.
- ***Metal Siding***: To maintain the modern, industrial aesthetic of container homes, metal sidings such as steel or aluminum can be used. These materials are durable, fire-resistant, and available in a variety of finishes and colors.

They reflect the original spirit of the container and can be installed with insulation to improve energy efficiency.

- ***Stucco***: Applying stucco is another way to transform the appearance of a container home. It can be textured or smooth and painted in any color, offering versatility in design. Stucco also adds an additional layer of insulation and protection against the elements.

- ***Composite Panels***: These are modern materials that offer high durability and low maintenance. Composite panels can mimic the appearance of wood, stone, or other textures, providing an aesthetic appeal without the upkeep of natural materials.

Beyond material selection, there are several techniques that can be employed to enhance the curb appeal of container homes:

- ***Color Strategy***: The color of the exterior plays a significant role in defining the home's character. Light colors can make the home appear larger and blend into a sunny landscape, while dark colors can offer a striking contrast against a lighter backdrop. Bright, bold colors can express individuality and make the home stand out.

- ***Architectural Features***: Adding architectural elements such as a pitched roof, decks, or pergolas can break up the monotony of the flat container walls and roofs. These features not only add to the visual interest but also improve the functionality of the space.

- ***Lighting***: Strategically placed outdoor lighting can enhance the home's architecture and make it inviting at night. It can accentuate architectural details, illuminate pathways, and add a layer of security.

The landscaping around the home should complement the exterior design and help integrate the container into its surroundings. Native plants, decorative grasses, and trees can soften the edges and help the home blend with the local environment.

Additionally, features like stone paths, water elements, and outdoor art can enhance the overall curb appeal and create a cohesive look.

Enhancing the curb appeal of a container home involves thoughtful consideration of exterior finishes and integration into the surrounding landscape. By selecting the right materials and employing various design techniques, homeowners can transform a basic container into a stunning and inviting residence. This not only makes the home a pleasant place to live but also increases its value and appeal in the housing market. The process of choosing and applying exterior finishes is an opportunity to express creativity and ensure that the home reflects the owner's style and values, all while contributing to its longevity and sustainability.

Creating Outdoor Living Spaces

Outdoor living spaces are integral to the modern lifestyle, offering a seamless blend of comfort and nature that enhances the overall living experience. For container homes, these spaces are not just aesthetic additions but functional extensions of the indoor environment, providing areas for relaxation, entertainment, and interaction with the natural world. This section explores how homeowners can create and optimize outdoor living spaces in container homes, emphasizing design strategies that enhance functionality and aesthetic appeal while fostering a connection to the outdoors.

The key to creating effective outdoor living spaces lies in understanding their intended use and ensuring a smooth transition between indoor and outdoor areas. Container homes, with their modular nature, offer unique opportunities for innovative outdoor space integration. The design process begins with a clear definition of the space's purpose, whether it's dining, lounging, cooking, or gardening. This definition guides the selection of features such as decks, patios, outdoor kitchens, and gardens.

Designing for functionality involves considering the flow of movement between spaces. For example, if outdoor dining is a priority, the space should be easily accessible from the kitchen. Similarly, lounge areas should be situated in quieter parts of the garden to provide a peaceful retreat. Effective design not only maximizes usability but also enhances the visual appeal of the outdoor space, making it an inviting extension of the home.

The choice of materials for outdoor living spaces in container homes should reflect both the aesthetic of the home and the durability needed to withstand the elements. Natural materials like stone and wood can be used to soften the industrial feel of the containers and help integrate the structure into the landscape. These materials can be used for decking, patio flooring, and as decorative elements throughout the garden.

Incorporating features like pergolas, canopies, or retractable awnings can add both functionality and style to outdoor spaces. These features provide shade on sunny days and protection during inclement weather, making the outdoor space usable year-round. Additionally, integrating outdoor fire pits or fireplaces can create a cozy atmosphere for evening gatherings, extending the usability of outdoor spaces into cooler seasons.

Landscaping is a critical component of outdoor living space design. The right plant selection and placement can enhance privacy, provide shade, and add color and texture to the garden. When selecting plants, homeowners should consider the local climate and the amount of maintenance they are willing to undertake. Native plants are often recommended because they require less water and care than non-native species and are better for the local ecology.

Creating themed gardens, such as xeriscapes or rain gardens, can add unique elements to the outdoor space while being environmentally responsible. Xeriscapes are particularly suitable for dry climates as they require minimal irrigation, while rain gardens can help manage stormwater runoff in wetter regions.

Furnishing outdoor living spaces should reflect the style and comfort of the home's interior. Durable, weather-resistant furniture that complements the home's design theme can create a cohesive look.

Incorporating outdoor rugs, cushions, and throws can add comfort and style, making the outdoor space more inviting. When choosing furnishings, consider materials that can withstand local weather conditions and provide ease of maintenance.
For those with limited space, creating multi-functional outdoor areas is essential. Features like foldable or multi-purpose furniture and vertical gardens can maximize space usage. Outdoor kitchens can serve multiple functions, providing a space for cooking, dining, and socializing. Similarly, installing a projector screen can turn a simple deck into an outdoor movie theater, offering another way to enjoy the space.
Creating outdoor living spaces in container homes is about more than just extending the living area; it's about enhancing quality of life by connecting residents with the outdoors. Through thoughtful design, selection of durable materials, strategic landscaping, and personal touches in furnishing, these spaces can transform how residents view and interact with their home environment. Outdoor living spaces not only increase the functional area of the home but also add significant value, both in terms of property value and lifestyle quality. By considering the flow, functionality, and aesthetic integration of these areas, container homeowners can craft spaces that are not only beautiful and useful but also harmonious extensions of their indoor living environments.

Eco-Friendly Landscaping and Permaculture Principles

Incorporating eco-friendly landscaping and permaculture principles into the design of container homes offers a unique opportunity to enhance not only their aesthetic and market value but also their environmental sustainability. This approach focuses on creating living spaces that are in harmony with nature, conserving resources, and supporting biodiversity. By understanding and applying these principles, homeowners can transform their properties into models of sustainability and self-sufficiency. Eco-friendly landscaping is an approach that reduces the environmental impact of a garden. It emphasizes the use of native plants, conservation of resources, and natural maintenance techniques.

This type of landscaping encourages homeowners to see their outdoor spaces as part of a larger ecosystem, interacting with the local environment in a supportive and sustainable way.

Permaculture is a design philosophy that seeks to create sustainable and self-sufficient landscapes based on natural ecosystems. It involves designing these landscapes to meet human needs while maintaining the health and balance of the natural environment. Permaculture is built on the idea of making the least change for the greatest possible effect, and it emphasizes the harmonious integration of the landscape with people and their homes.

Designing a landscape with permaculture principles starts with careful observation and interaction with the current environment. This might include noting the sun's path, existing vegetation, and natural water flow. With this information, a design can be developed that utilizes the landscape's natural tendencies and advantages. For example, in permaculture, every element is designed to perform multiple functions. A tree might provide shade, fruit, support for climbing plants, and habitat for birds, all at once.

In regions where water is scarce, the choice of plants and landscape design can significantly affect water usage. Using drought-tolerant or native plants can drastically reduce the need for watering. Techniques such as rainwater harvesting, which involves collecting rain from rooftops and storing it for garden use, and mulching, which reduces water evaporation from the soil, are both effective ways to manage landscape hydration without relying heavily on municipal water sources.

Healthy soil is crucial for a productive garden. Permaculture advocates for enhancing soil health naturally through composting organic waste and employing no-dig gardening techniques. Composting reduces waste and enriches the soil, while no-dig gardening preserves soil structure and encourages a vibrant, healthy microbiome. These practices support robust plant growth and reduce the need for chemical fertilizers. Permaculture landscapes use natural methods for managing pests. This includes companion planting, where certain plants are known to enhance each other's growth or deter pests naturally.

Also, encouraging beneficial insects by creating habitats can help maintain a natural balance, reducing the need for chemical pesticides.

The principles of eco-friendly landscaping and permaculture can be seamlessly integrated into outdoor living areas. Designing spaces that blend the indoors with the outdoors enhances the usability and enjoyment of these areas. For instance, incorporating an edible garden not only provides fresh produce but also brings life and energy to the space. Designing these areas with sustainability in mind ensures that they are beautiful, functional, and environmentally responsible.

By integrating eco-friendly landscaping and permaculture principles into their design, container homes can become beacons of sustainability. These principles allow homeowners to create outdoor spaces that are not only visually appealing and functional but also have a positive impact on the environment. They turn simple gardens into ecosystems that support local wildlife, conserve resources, and provide a natural escape for their inhabitants. Through careful planning and thoughtful design, container homeowners can forge a deep connection with the natural world, right in their own backyards.

Conference room

Chapter 12: Navigating the Permit Process and Hiring Contractors

Step-by-Step Guide to Obtaining Permits

Making a container home involves navigating a series of regulatory hurdles, including obtaining the necessary permits. This process can be daunting, especially for first-time builders, but understanding the steps involved can make it more manageable. This section offers a comprehensive guide to the permit process for building a container home, detailing what to expect, how to prepare, and how to ensure compliance with all local zoning and building codes.

Before embarking on the journey of building a container home, it's crucial to understand the local building codes and zoning regulations that apply to your property. These regulations are designed to ensure that all structures are safe, compliant with environmental standards, and in harmony with the surrounding community. The first step in the permit process is to visit your local planning or building department to gather all necessary information regarding the zoning laws, building codes, and any restrictions that might affect your project.

One of the most critical aspects of the permit application process is the submission of detailed plans of the proposed home. These plans should include site layouts, floor plans, elevations, and details about utilities and other systems. The plans must be detailed enough to show compliance with local building codes, including structural integrity, fire safety, electrical systems, plumbing, and accessibility. It may be necessary to hire an architect or engineer to prepare these plans, especially if the local jurisdiction has specific requirements for professional endorsements.

Engaging early and often with local authorities can facilitate a smoother permit process. This might involve preliminary meetings with planning officers to discuss your project before submitting formal applications. These discussions can provide valuable insights into potential challenges or additional requirements that could arise during the review process.

Being proactive and establishing a good rapport with local officials can also expedite the approval process and help avoid costly revisions.

Once your plans are ready and you have a clear understanding of local regulations, the next step is to submit your permit application. This application should include all required documents, plans, and fees. It's important to ensure that the submission is as complete as possible to avoid delays. After submission, there will be a review period during which local officials will scrutinize the plans to ensure they meet all legal and safety standards.

The review process can vary significantly depending on the locality. It may involve several rounds of reviews and requests for additional information. During this phase, the building department may require changes to the plans to comply with code requirements. It's important to respond to these requests promptly and to maintain flexibility in your design to accommodate necessary changes. This part of the process can be iterative, requiring patience and a willingness to collaborate closely with the authorities.

It's not uncommon to face challenges and obstacles during the permit process. These could range from simple misunderstandings about building requirements to more significant disputes about zoning or code interpretations. In such cases, it's essential to remain professional and cooperative, seeking to understand the concerns of the authorities and finding constructive solutions. Sometimes, it may also be necessary to seek legal advice or the services of a planning professional to navigate complex regulatory challenges.

Once all requirements are met and your plans are approved, you'll receive your building permits. However, obtaining permits is not the end of the regulatory process. You must also prepare for several inspections throughout the construction phase to ensure ongoing compliance with building codes and permit conditions. These inspections are crucial as they verify that the construction matches the approved plans and adheres to all safety standards.

Navigating the permit process for building a container home requires thorough preparation, a clear understanding of local regulations, and effective communication with local authorities.

By carefully planning and engaging with the permitting process, prospective builders can ensure that their project proceeds smoothly and remains compliant with all necessary regulations. This approach not only facilitates the successful completion of the project but also ensures the safety and durability of the container home.

Tips for Selecting and Working with Contractors

Building a container home often involves collaborating with various contractors, from general contractors who oversee the entire project to specialized tradespeople like electricians and plumbers. The success of your container home project can hinge significantly on your ability to choose the right contractors and manage these relationships effectively. This comprehensive guide explores the nuances of selecting contractors, establishing clear communication, and ensuring a productive working relationship throughout the construction of your container home.

Understanding the Role of Contractors in Building Container Homes

Contractors play a critical role in translating your container home dreams into reality. They possess the expertise and resources necessary to handle the complexities of construction, from ensuring compliance with local building codes to executing the detailed design elements of your home. A general contractor typically manages the overall project, coordinating with specialized subcontractors to complete specific tasks.

Finding a contractor who is experienced with container homes can add immense value. Given the unique challenges and techniques involved in building with shipping containers, experience in this area can lead to better problem-solving on-site and more effective execution overall.

1.Research: Start by conducting thorough research. Look for contractors with specific experience in container construction. You can find potential contractors through online searches, recommendations from within the container home community, or by consulting local building associations.

2.Check Credentials and Reviews: Once you have a list of potential contractors, check their credentials. Ensure they are licensed, insured, and have a good standing with building authorities. Online reviews and references from past clients can provide insight into their reliability, quality of work, and customer service.

3.Interview Prospective Contractors: Interviewing contractors gives you a chance to ask detailed questions about their experience with container homes, understand their approach to potential challenges, and gauge their enthusiasm for your project. This is also the time to discuss their availability and get a sense of their communication style.

A well-defined contract is fundamental to managing your relationship with a contractor. It should clearly outline the scope of work, budget, timelines, payment schedules, and other critical project details.

1.Scope of Work: Define all tasks that the contractor will perform, including any expected standards for the materials and workmanship. The more detailed your project scope, the less room there is for misunderstandings.

2.Budget and Payment Terms: The contract should clearly state the total cost of the construction and set terms for payments. This might include a down payment, progress payments, and final payment upon completion. Be sure to discuss how potential changes in work will be handled to avoid unexpected costs.

3.Timeline: Include a detailed timeline with milestones for specific aspects of the project. This helps in tracking progress and ensures that both you and the contractor are aligned on expectations.

Maintaining open, honest, and regular communication with your contractor is vital. Regular meetings, whether in person or via digital platforms, can help keep everyone on the same page and address issues as they arise.

1.Regular Updates: Agree on a schedule for regular updates and stick to it. During these updates, discuss the progress of the project, any challenges encountered, and any adjustments needed to the original plan.

2.Be Available: Be available to make decisions quickly. Delays in decision-making can lead to project delays and increased costs.

3.Keep Records: Keep detailed records of all communications, decisions, and changes related to your project. This documentation can be invaluable in resolving any disputes and for your own reference throughout the project.
Even with the best planning, challenges can arise. Whether it's unexpected construction issues or budget overruns, how these challenges are managed can affect the outcome of your project.

1.Anticipate and Plan: Anticipate potential issues and discuss these scenarios with your contractor in advance. Understanding how unforeseen circumstances will be handled before they occur can help mitigate stress and confusion.

2.Dispute Resolution: Include a dispute resolution process in your contract. Methods such as mediation or arbitration can provide a way to resolve disputes without resorting to litigation.
Selecting the right contractors and maintaining a good working relationship with them are crucial to the successful completion of a container home. By taking the time to carefully choose contractors, set clear expectations in a well-drafted contract, and communicate effectively throughout the project, you can help ensure that your container home is built to your satisfaction while also managing to stay within your planned budget and timeline. This careful preparation and partnership can transform the construction process from a potential source of stress into an exciting journey towards creating your dream home.

Building Inspections and Compliance

Navigating the complex landscape of building inspections and ensuring compliance with local building codes is crucial for constructing a container home. This detailed exploration covers the importance of adhering to building regulations, preparing for inspections, and the role these inspections play in maintaining safety and legal standards. Compliance not only ensures the structural integrity and safety of your home but also impacts its long-term viability and legality.

Building codes are set by local, state, or national authorities to establish minimum standards for construction projects. These standards cover safety, health, and environmental compliance. For container homes, which may not fit traditional building models, it is crucial to understand specific codes that apply. These codes encompass a wide range of construction aspects including electrical systems, plumbing, fire safety, and structural integrity. The first step in ensuring compliance is to thoroughly research and understand the building codes applicable in your area, typically available through local government websites or local building authority offices.

The Importance of Compliance

Compliance with building codes is designed to protect the health and safety of occupants, ensuring that the structure can withstand environmental stresses, electrical systems are safe, and plumbing ensures proper sanitation. Non-compliance can lead to fines, legal action, and the necessity to redo non-compliant work, which can significantly increase both the cost and timeline of your project. Additionally, a compliant home is easier to sell and can be insured at standard rates, whereas non-compliance can severely affect both the resale value of your home and the ability to obtain insurance. Building inspections are typically conducted at multiple stages of the construction process. Preparing adequately for each inspection is vital for the smooth progression of your building project.

The inspection process starts with checking the site before the foundation is poured, ensuring it is set up correctly according to plans.

As the frame of the house goes up, an inspection is conducted to look at the structural integrity and proper installation of systems before walls are closed in. Once construction is completed, a final inspection ensures that the building is fully compliant and ready for occupancy.

Strategies for Passing Inspections

Preparation is key before an inspection. Review your construction against the approved plans and building codes and correct any issues beforehand to avoid failing the inspection. Keeping detailed records of all construction processes, materials used, and compliance checks can help resolve any questions the inspector might have. Establishing a good rapport with inspectors can also facilitate a smoother inspection process. Be respectful, open to learning, and ready to make necessary changes suggested by the inspectors.

If an inspection is failed, it is crucial to view this as an opportunity to improve. Inspectors provide detailed reports on what aspects of the construction do not meet code and offer advice on how to rectify these issues. Understanding the issues clearly is essential if necessary, seek further explanation or a demonstration from the inspector. Make the required adjustments, possibly hiring necessary tradespeople to correct the issues. After corrections are made, a re-inspection ensures everything is now up to code. This process continues until final approval is received. Handling building inspections and ensuring compliance are integral and complex parts of constructing a container home. By understanding and adhering to relevant building codes, preparing diligently for inspections, and maintaining open communication with inspectors, homeowners can ensure their container home is safe, compliant, and built to last.

This meticulous adherence not only secures the structural integrity and legality of the home but also enhances its longevity and value, making it a secure investment for the future.

Chapter 13: Budget Management and Cost-Saving Strategies

Detailed Breakdown of Potential Costs

Producing a container home involves various costs, some obvious and some hidden. Understanding these costs in detail is crucial for effective budget management. This comprehensive analysis explores all potential expenses associated with constructing a container home, providing future homeowners with the insights needed to plan and control their budgets effectively.

The first cost in building a container home is the purchase of the containers themselves. Prices can vary widely based on the size, condition, and history of the containers. New or "one-trip" containers are more expensive but offer structural integrity and fewer cosmetic issues. Used containers can be significantly cheaper but might require extensive modifications and repairs, which can add to the total cost.

Once the containers are purchased, they must be prepared for habitation. This involves cleaning, removing any hazardous materials used in shipping, and making structural modifications such as cutting openings for doors and windows. These initial modifications require professional equipment and expertise, adding to the preparation costs.

Before a container can be placed, the site must be prepared. This includes clearing the land, leveling the area, and potentially grading it for drainage. Each of these steps involves labor and machinery, which can be costly depending on the land's initial condition.

The foundation is another significant expense. Container homes need a solid foundation just like any other home. Options range from simple concrete piers to more complex slab foundations, depending on the design and local building codes. The foundation not only supports the structure but also helps to protect the containers from corrosion and moisture.

Connecting a container home to utilities is another area where costs can accumulate. This includes running water, electricity, sewage, and possibly gas. Each utility may require a separate installation and connection fee. In remote locations, these costs can increase substantially, especially if the site requires significant work to access municipal services or if alternative systems like septic tanks or solar panels are needed.

Proper insulation is critical in container homes to make them livable. The cost of insulation varies depending on the materials used and the climate where the home is located. Interior finishing, including drywall, flooring, and painting, also requires a significant budget allocation. These finishes are not just cosmetic; they contribute to the home's functionality and comfort.

Exterior finishes can help to protect the container against the elements and improve its appearance. Options include cladding, painting, or adding a roof over the containers. Each option comes with different costs and benefits in terms of durability and maintenance. Modifications might also include structural reinforcements to ensure the home can withstand local weather conditions and environmental stresses.

Labor is one of the most significant costs in building a container home. Skilled labor is necessary for tasks such as welding, carpentry, and electrical work. The cost of labor can vary widely depending on the region, the complexity of the project, and the availability of skilled professionals.

Permits and Legal Fees

Obtaining the necessary permits can be both time-consuming and expensive. The cost includes not only the fees for the permits themselves but also any legal advice or services needed to navigate the local building regulations.

Finally, once the container home is built, it needs to be furnished. The cost of furniture and decor can vary greatly depending on tastes and preferences. While some opt for high-end, custom pieces, others find savings by choosing second-hand or multipurpose furniture.

Understanding the detailed breakdown of potential costs involved in building a container home is essential for anyone considering this type of construction. By anticipating these costs and planning accordingly, prospective homeowners can manage their budgets more effectively, ensuring that their container home project is both financially viable and successful. This level of detailed planning helps avoid unexpected expenses and ensures that the home meets all expectations both structurally and aesthetically.

How to Save Money Without Compromising Quality

Producing a container home offers a unique opportunity to create a personalized living space while potentially saving money on construction costs. However, balancing budget constraints with the desire for a high-quality home requires strategic planning and smart decision-making. This detailed guide explores various techniques for reducing costs without sacrificing the quality and durability of your container home, ensuring that every dollar spent adds value and functionality.

The foundation of cost-saving in container home construction lies in meticulous planning and design. By carefully considering each aspect of your home's layout and functionality in the planning stages, you can avoid costly changes and revisions during construction.

- ***Optimize Space Usage***: Container homes naturally encourage space efficiency due to their compact size. Design your space to maximize usability without unnecessary expansions that increase costs. Multi-functional furniture and innovative storage solutions can enhance living areas without the need for additional containers.

- ***Simple Design Choices***: Simplifying the design of your home can lead to significant savings. Complex designs often require more specialized labor and materials. A minimalist approach not only reduces these costs but also aligns with the modern aesthetic often associated with container homes.

Choosing the right materials can have a substantial impact on the overall cost of building a container home.

- ***Recycled and Reclaimed Materials***: Utilizing recycled or reclaimed materials can reduce costs substantially. Many building materials, such as wood, glass, and metal, can be sourced from salvage yards or through online marketplaces from other renovation projects.

- ***Local Sourcing***: Purchasing materials from local suppliers can save money on transportation costs and support the local economy. Additionally, local materials are often better suited to the regional climate and environmental conditions, enhancing the sustainability of your home.

- ***Quality vs. Cost***: Invest in quality where it matters. Durable materials might be more expensive upfront but can save money in the long run by reducing maintenance and replacement costs. Identify which aspects of your home are worth investing in and where you can opt for less expensive alternatives without compromising the structure's integrity.

While DIY approaches can save costs, using professional services strategically can prevent expensive mistakes and ensure that the work is done right the first time.

- ***Hire Skilled Labor for Critical Tasks***: For elements like electrical work, plumbing, and structural modifications, hiring experienced professionals can prevent costly errors and compliance issues.

- ***Manage Some Tasks Yourself***: Tasks such as painting, installing insulation, or finishing interiors can be managed personally if you have the time and skills. This hands-on approach can significantly reduce labor costs.

Innovative construction methods can not only save money but also enhance the efficiency and sustainability of your container home.

- ***Prefabrication***: Where possible, use prefabricated components. These are often more cost-effective and result in quicker assembly on site, reducing labor costs and construction time.

- ***Energy Efficiency***: Investing in energy-efficient windows, insulation, and appliances can reduce long-term operating costs. Although the initial investment might be higher, the savings on utility bills will benefit your finances in the long run. Being a savvy shopper can help in managing construction costs effectively.

- ***Bulk Purchases***: If possible, buy materials in bulk. Bulk purchasing can often secure discounts, especially for standard items like screws, nails, insulation materials, and paint.

- ***Negotiate Prices***: Don't hesitate to negotiate prices with suppliers and contractors. Especially for larger orders or long-term relationships, suppliers are often willing to offer discounts to secure the business.

Maintaining a close watch on your expenses and keeping detailed records throughout the construction process can help you stay within budget.

- ***Track All Costs***: Keep detailed accounts of all expenses, including receipts, invoices, and statements. This will help you understand where your money is going and identify areas where you can cut costs without compromising quality.

- ***Review and Adjust Budgets Regularly***: Regularly review your spending against your budget. If you notice you're spending more than planned, reassess your choices and adjust your plans to ensure you can complete your project without financial strain.

Building a container home on a budget does not mean you have to compromise on quality. By planning carefully, making informed choices about materials and labor, and managing the construction process diligently, you can create a cost-effective home that is both beautiful and durable. The key is to prioritize expenses, invest in quality where it matters, and continuously seek out opportunities to reduce costs without undermining the integrity and functionality of your home. With the right strategies, your container home can be a model of both financial prudence and excellent design.

Financing Your Container Home Project

Embarking on the journey of building a container home is not only a creative architectural endeavor but also a significant financial investment. For many prospective homeowners, understanding and securing the right financing is a crucial step in making their dream home a reality. This in-depth exploration covers various financing options available for container home projects, strategies to secure funding, and tips on managing financial resources effectively throughout the construction process.

Before diving into financing options, it is essential to have a clear understanding of the overall financial requirements of building a container home. This includes not only the cost of the containers themselves but also expenses related to site preparation, construction, utilities, interior and exterior finishes, and any unforeseen costs. Accurately estimating these costs will provide a solid foundation for securing financing.

Many individuals choose to use personal savings to finance their container home projects. This method has the advantage of not relying on external lenders and avoiding interest costs, providing a significant degree of financial independence. However, using personal savings requires disciplined saving and budgeting strategies to accumulate the necessary funds without compromising other financial goals.

For those who do not have sufficient savings, construction loans can be an effective way to finance a container home. These loans are specifically designed for financing home construction projects and are usually disbursed in stages as the construction progresses. Lenders typically require detailed construction plans and a projected budget before approving a construction loan. The terms and interest rates of construction loans can vary significantly, so it's crucial to shop around and compare offers from different lenders.

Securing a mortgage for a container home can be more challenging than for a traditional home due to the unconventional nature of container construction. However, more lenders are becoming open to financing alternative housing projects as the popularity of container homes grows. To improve your chances of securing a mortgage, you may need to provide comprehensive documentation demonstrating the expected stability and value of the property upon completion.

In some regions, government programs may offer loans or grants specifically for building alternative or eco-friendly homes. These programs can provide favorable terms, such as low-interest rates or no repayment obligations for grants. Researching local government programs and eligibility requirements can uncover potential funding opportunities that could significantly reduce the financial burden of your project.

Another creative financing option is seeking funding from private investors or launching a crowdfunding campaign. These methods can be particularly viable if your container home project has a unique appeal or environmental benefits that resonate with potential investors or the community. Crafting a compelling story and demonstrating the project's potential value are crucial for attracting investment through these channels.

Tips for Securing Financing

Securing financing requires careful preparation and a strategic approach. Here are some tips to improve your chances of obtaining funding:

- ***Strong Business Case***: Treat your container home project like a business venture when applying for financing. Prepare a detailed business plan that includes market analysis, detailed budgets, timelines, and potential return on investment.

- ***Good Credit Score***: A high credit score can significantly enhance your ability to secure favorable loan terms. Before applying for financing, check your credit score and take steps to improve it if necessary.

- ***Cost-effective Design***: Design your container home in a way that maximizes cost efficiency without compromising on essential features. A cost-effective design can make the project more attractive to lenders by reducing the required loan amount.

- ***Professional Support***: Engaging professionals such as architects, engineers, and financial advisors can lend credibility to your project. Professional endorsements can reassure lenders of the viability and potential success of your container home project.

Financing a container home project involves navigating a complex landscape of personal savings, loans, and possibly external investment. Understanding the full scope of financial requirements, exploring diverse financing options, and preparing a solid application are key to securing the funds necessary to bring your container home to life. With careful planning and strategic financial management, building your dream container home can become an achievable reality, paving the way for a unique and personally rewarding living space.

Chapter 14: Common Pitfalls and How to Avoid Them

Lessons Learned from Real Container Home Projects

Making a container home is an innovative approach to housing that combines architectural creativity with practical challenges. Those who have ventured into this realm have accumulated a wealth of knowledge that is invaluable for anyone considering a similar path. This comprehensive exploration delves into the collective experiences of container home builders, highlighting the essential lessons learned and the typical pitfalls encountered throughout the construction process.

Thorough planning and research are critical from the outset of a container home project. Understanding the scope and requirements goes beyond simple design and encompasses legal, environmental, and practical considerations. One of the most common mistakes is failing to adequately research local zoning laws and building codes. Many builders have faced setbacks after realizing their projects did not comply with local regulations, leading to costly modifications or even project termination. Engaging with local planning offices early in the process is essential to integrate all legal requirements into the project planning effectively.

Selecting the right containers is also crucial. Not all shipping containers are suitable for building. Factors such as the age of the container, its previous usage, and structural integrity are critical. Builders recommend purchasing containers from reputable suppliers and inspecting them in person before purchase to avoid issues related to poor structural quality, which can escalate costs and complexities later on. Adapting shipping containers into livable spaces involves significant structural modifications, which can be complex and fraught with potential mistakes. Cutting steel for doors, windows, and room openings without proper reinforcement can compromise the container's integrity.

Experienced builders emphasize the importance of consulting with structural engineers to ensure all modifications meet safety standards and that the structure is adequately reinforced.

Proper insulation is vital to manage interior temperatures and prevent condensation issues. Builders who have skimped on this aspect often end up with homes that are too hot in summer and too cold in winter, leading to uncomfortable living conditions and higher energy costs. Using high-quality insulation materials and techniques suited to the local climate is crucial.

Installing utilities in a container home can present unique challenges, especially for those unfamiliar with the nuances of plumbing and electrical layouts in metal structures. Improper installation of electrical and plumbing systems can lead to long-term problems like electrical shorts or water leaks. Hiring qualified professionals who have experience with container structures is essential to ensure that all systems are safely and effectively installed.

Maximizing space in container homes requires innovative interior design solutions. Custom solutions and multi-functional furniture are important to enhance livability and storage without cluttering the space, maintaining a balance between functionality and aesthetic appeal.

Given their steel composition, containers are susceptible to rust and corrosion, particularly in humid or saline environments. Applying protective coatings and performing regular maintenance checks are recommended to preserve the container's exterior and ensure its longevity.

The importance of a solid foundation cannot be overstated. Inadequate foundations have led to settling and structural issues in some container homes. Preparing the site correctly and choosing a foundation type that suits the soil, climate, and topography is vital.

The journey of building a container home is replete with unique challenges and opportunities for learning.

By absorbing the lessons learned by those who have already embarked on this path, prospective builders can avoid common pitfalls and approach their projects with a greater understanding of what it takes to create a successful container home.

This knowledge aids in avoiding costly mistakes and helps achieve a more sustainable, efficient, and personalized living environment, refining the art and science of container home construction and contributing to a broader knowledge base that supports this innovative approach to modern living.

Addressing Challenges and Solutions

Building a container home presents a set of unique challenges, each demanding innovative solutions and careful consideration. This exploration delves into common issues faced during the construction of container homes and offers strategic advice on how to address these challenges effectively, ensuring a smooth and successful project completion.

One of the primary challenges in container home construction is maintaining structural integrity while making necessary modifications. Cutting and modifying containers for doors, windows, and other structural openings can compromise their strength.

The key to overcoming this challenge lies in expert consultation. Hiring a structural engineer who understands the dynamics of shipping container modifications can provide invaluable insights. Engineers can help design appropriate reinforcement strategies that ensure safety without compromising the container's structural integrity. Utilizing computer-aided design (CAD) software to visualize modifications and potential structural reinforcements before actual cutting can save both time and resources.

Container homes are prone to extreme thermal conditions because metal conducts heat and cold efficiently. Without proper insulation, container homes can become uncomfortably hot in summer and freezing in winter.

To tackle this, effective insulation is crucial. Spray foam insulation, while more expensive, offers excellent thermal resistance and adds structural rigidity to the container walls. Additionally, installing a reflective roof coating or green roof can deflect sunlight and improve the overall thermal performance of the home.

Integrating passive solar design principles can also capitalize on the environmental conditions to maximize heating and cooling efficiency.
Moisture accumulation and condensation are significant challenges in container homes due to the metal walls and often compact design.
Installing adequate ventilation systems is essential to prevent moisture buildup. This includes both active systems, like exhaust fans in kitchens and bathrooms, and passive systems, such as strategically placed vents that facilitate natural air flow. Additionally, using moisture barriers and ensuring that insulation materials are properly sealed can prevent condensation from penetrating structural elements.
Navigating zoning laws and building regulations can be a major hurdle due to the non-traditional nature of container homes. Many localities may not have specific codes addressing container homes, leading to potential regulatory challenges.
Engaging with local building officials early in the project can help identify and solve compliance issues before they become significant obstacles. It's advisable to present detailed plans and be prepared to explain how the project meets or exceeds building codes. In some cases, securing the services of an advocate experienced in alternative housing can facilitate discussions with zoning boards and regulatory bodies.
Securing financing and insurance for a container home can be challenging, as many financial institutions and insurance companies are unfamiliar with the container housing concept.
Providing lenders and insurers with detailed documentation about the construction process, safety features, and compliance with building codes can help mitigate their concerns. It may also be beneficial to present case studies or examples of successfully financed and insured container homes. Establishing a clear valuation for the home, backed by appraisals from certified professionals, can further reassure financial institutions and insurance companies.
Integrating a container home into its surrounding environment and ensuring it meets the owner's aesthetic expectations requires careful design and customization. Working with architects and designers who have experience with container homes can ensure that the home is not only functional but also aesthetically pleasing.

Incorporating natural elements, such as wood cladding, or using landscaping effectively can help the home blend with its surroundings and enhance its curb appeal.

Building a container home involves navigating a complex array of challenges, from structural and thermal issues to regulatory and financial hurdles. Each challenge requires a thoughtful solution, often involving professional expertise and proactive planning. By addressing these challenges head-on with informed strategies and expert guidance, builders and homeowners can significantly increase their project's success rate, resulting in a durable, comfortable, and beautiful container home.

Maintenance and Upkeep Tips

Maintaining a container home efficiently requires a unique approach compared to traditional homes, given the distinct materials and construction techniques used. Proper maintenance ensures the longevity and sustainability of a container home, protecting the investment and keeping it safe and comfortable for its inhabitants. This extensive guide delves into the essential maintenance and upkeep practices specifically tailored for container homes, highlighting the best strategies and solutions to preserve these unique structures over time.

Container homes, predominantly made of steel, face specific challenges such as corrosion, insulation performance, and structural integrity over time. Unlike traditional homes, the steel structure of a container can be susceptible to rust and degradation if not properly maintained. Moreover, the compact nature of many container homes can increase the wear and tear on its components due to the denser living conditions.

The key to effective maintenance is the regular inspection of the container home, focusing on its critical aspects:

- ***Rust and Corrosion Prevention***: Regularly inspect the exterior and interior of the container for signs of rust or corrosion. Areas where the paint has chipped away or metal is exposed are particularly vulnerable. Applying a rust-inhibiting paint or sealant as soon as any signs of rust appear can prevent further degradation.
It's advisable to perform this inspection at least twice a year or more frequently in harsh climates.

- ***Seal Integrity***: Check the seals around windows, doors, and joints regularly for leaks or drafts. Container homes can shift slightly with temperature changes, which can cause seals to crack or loosen. Reapplying sealant and replacing weather-stripping as needed can help maintain temperature control and prevent water damage.

- ***Roof and Drainage Systems***: Ensure that the roof and gutter systems are free from debris that could cause water to pool and lead to rust or leakage. The roof of a container home should have adequate water redirection systems to avoid accumulation, which could lead to serious structural issues.
Insulation and ventilation are crucial in maintaining the comfort and air quality inside a container home. Over time, the performance of insulation materials can degrade, or they might be compromised by moisture or pest infestations.

- ***Inspect Insulation Regularly***: Check the condition of the insulation regularly, particularly if changes in interior climate control are noticed. Look for signs of dampness, mold, or pest activity, which can indicate that the insulation may need to be replaced or supplemented.

- ***Ventilation System Upkeep***: Ventilation systems in a container home should be checked to ensure they are clear of blockages and functioning correctly. Proper ventilation is crucial to prevent condensation, which can lead to rust and mold growth.

The unique construction of container homes can sometimes lead to challenges with plumbing and electrical systems, especially if modifications were made during the home's conversion.

- ***Regular Plumbing Checks***: Given the compact nature of many container homes, plumbing issues can quickly escalate. Regular checks for leaks, water pressure problems, or drainage issues can prevent minor problems from becoming major repairs.

- ***Electrical Safety Inspections***: Electrical systems should be inspected periodically by a qualified electrician to ensure they are safe and meet any updated regulations. This is particularly important if the container home features modifications or DIY electrical work.

Maintaining the appearance of a container home not only involves structural and functional considerations but also aesthetic upkeep.

- ***Exterior Painting***: The exterior paint on a container home does more than just appeal; it acts as a protective layer against the elements. Regularly cleaning the exterior and applying fresh paint when signs of wear appear can protect against rust and corrosion.

- ***Interior Touch-ups***: Inside the home, maintaining wall surfaces, floors, and fixtures can keep the home looking fresh and new. Regular cleaning, along with periodic updates to the interior décor, can also help manage wear and tear.

Effective maintenance and upkeep of a container home are crucial to preserving its value and functionality over time. By conducting regular inspections, addressing issues promptly, and understanding the unique needs of a container structure, homeowners can ensure their home remains a safe, comfortable, and aesthetically pleasing space for years to come. With the right care, container homes can defy their humble beginnings to become enduring and sustainable living spaces.

Chapter 15: Case Studies and Success Stories

Inspirational Stories from Container Home Owners

The journey to build and live in a container home is filled with challenges and triumphs, and each container home has a unique story to tell. These stories not only inspire but also educate and guide new builders through the labyrinth of creating a sustainable and personalized living space out of shipping containers. This chapter explores several inspirational stories from container home owners who have transformed their dreams into reality, providing insights into the creative processes, obstacles overcome, and the profound impacts these homes have had on their lives.

The Meadows family, consisting of Jim and Clara and their two children, embarked on their container home project in rural Montana. The family, driven by a desire for a sustainable lifestyle and closer connection to nature, decided to build their dream home using four high-cube shipping containers. Their journey began with a clear vision: to create a home that blended seamlessly with the surrounding landscape while maintaining a minimal environmental footprint.

The design phase was marked by a focus on large, floor-to-ceiling windows that not only invited ample natural light but also offered panoramic views of the majestic landscape. The family worked closely with an architect who specialized in container homes to ensure that their vision was structurally sound and feasible. One of the main challenges they faced was the local zoning regulations, which were initially not favorable to container constructions. However, with persistent efforts, detailed presentations to the planning boards, and modifications to their initial plans, they were able to gain the necessary approvals.

The construction phase was a learning curve for the Meadows. They decided to manage the project themselves, coordinating between various contractors and learning on the fly.

The family’s dedication to sustainability was reflected in their material choices, opting for recycled or sustainably sourced products wherever possible. Despite facing unexpected delays due to weather and some budget overruns, the family kept their spirits high and their vision clear.

The result was a stunning two-story home that embodied everything the Meadows had dreamed of. The containers were positioned to form a spacious open-plan living area in the center, flanked by private quarters on the sides. Innovative insulation techniques were employed to keep the home warm during harsh Montana winters and cool in the summer. The Meadows' story is not just about building a house; it's about creating a sustainable home that respects and enhances its natural surroundings.

Ellen, a retired school teacher, chose to downsize after her children moved out and decided that a container home in her beloved city of Austin, Texas, would be her next big project. Her goal was to construct a home that was both environmentally friendly and close to the urban core, allowing her to enjoy her retirement surrounded by the vibrant city life.

Ellen faced numerous obstacles, from finding a suitable plot within her budget to navigating the city's stringent building codes. She partnered with a local container home construction company that helped her design a compact, efficient home using two 40-foot containers. The design included a rooftop deck to provide outdoor living space, which was important to Ellen.

The most significant challenge was the limited space for transporting and placing the containers. Every step of the delivery and placement had to be meticulously planned. Despite these challenges, the project sparked interest and support from the community, many of whom were intrigued by the concept of container homes.

Ellen’s home turned out to be a marvel of modern design and functionality. Space-saving solutions such as built-in furniture and multi-purpose areas maximized her living space, proving that small-scale living did not require sacrificing comfort or style. Her story is a testament to the idea that personal challenges can be transformed into opportunities for growth and innovation.

These stories from real container home owners illuminate the path for future builders, showing that while the journey may be fraught with challenges, the rewards of perseverance, creativity, and commitment to sustainability are immense. Each story not only stands as a testament to personal achievement but also serves as a beacon of inspiration for those dreaming of building their own container homes. By sharing these experiences, potential builders gain not just the vision but also the practical knowledge needed to navigate their own construction challenges, transforming their own visions into reality.

Before and After Transformations

The transformation of simple shipping containers into fully functional and aesthetically pleasing homes is nothing short of remarkable. The journey from a raw, industrial container to a warm, inviting home encapsulates not only a physical metamorphosis but also a change in the lives of the homeowners. This exploration dives deep into the before and after transformations of several container home projects, illustrating the dramatic changes and the impact on the homeowners' lives.

Case Study 1: The Thompson Family Retreat

Before:

The Thompson family embarked on their container home project with three used shipping containers they acquired at a local port. Originally, these containers were slightly rusted, dented, and had remnants of their previous life carrying goods across the oceans. The family's plot of land was a bare piece of property in the outskirts of Nashville, Tennessee, overgrown with vegetation and lacking any development.

Transformation:

The transformation process began with the preparation of the land, clearing out the overgrowth and leveling the ground for a foundation. The containers underwent extensive modifications, including sandblasting to remove rust, applying corrosion-resistant paint, and cutting openings for windows and doors. A local architect helped design a layout that optimized the scenic views and natural light, creating a spacious living area by joining the three containers in a U-shape.

After:

The completed home is a testament to modern design and sustainability. The exterior features a combination of wood cladding and preserved original container walls, showcasing a rustic yet contemporary look. Inside, the home is outfitted with modern amenities, energy-efficient appliances, and stylish decor that reflects the family's tastes. The large glass doors open to a wooden deck that overlooks a newly landscaped garden, creating a seamless indoor-outdoor living environment.

Case Study 2: Alex's Downtown Studio

Before:

Alex, a young graphic designer, purchased a single 20-foot shipping container with the vision of creating a compact, efficient studio in the heart of Denver, Colorado. The container was in decent condition but required cleaning and minor repairs. The urban lot where Alex planned to place the home was small and constrained by its urban setting, surrounded by high buildings and busy streets.

Transformation:

The key challenge was to convert the small space into a functional and creative studio while adhering to the strict building codes of an urban area. The container was insulated with high-tech foam to combat the noise from the streets and maintain thermal comfort. Innovative space-saving furniture was custom-made to enhance the functionality of the studio. Solar panels were installed on the rooftop to maximize energy independence.

After:

Alex's studio is now a showcase of urban container home living, featuring a minimalist design with a focus on functionality. The interior sports a modern, monochromatic theme punctuated by bursts of color from Alex's artwork. Despite its size, the studio feels spacious due to clever design solutions like a murphy bed and a foldable work desk. The rooftop terrace offers an oasis in the city, with container gardening and solar-powered lighting.

Case Study 3: The Greenfield Eco-Farm

Before:

The Greenfields, passionate about sustainable living and permaculture, started with five containers placed on a rural piece of land in Oregon. The containers were aged and showed signs of extensive use but were structurally sound. The surrounding land was fertile but unkempt, with great potential for development into a sustainable farm.

Transformation:

The transformation involved extensive modifications to the containers to create a farmhouse that would serve both as a home and as an operational base for farming activities. Large cut-outs were made to install triple-paneled glass windows, and a green roof was added for better insulation and to reduce runoff. The interior design focused on using recycled materials, and a rainwater harvesting system was integrated into the home's design.

After:

The Greenfield Eco-Farm has become a model for sustainable agricultural practices and residential living. The farmhouse connects seamlessly with its surroundings, featuring exterior living walls that blend into the landscape. Inside, the home is equipped with state-of-the-art sustainable technologies, including a greywater recycling system. The farm produces a variety of organic crops, and the family hosts workshops on sustainable living, fulfilling their dream of contributing positively to their community.

These before and after stories of container homes demonstrate the immense potential of shipping containers as sustainable building materials. Each transformation story is unique, highlighting not just the physical renovation of spaces but also the positive changes in the lifestyles of the homeowners. Through creativity, dedication, and a focus on sustainability, these container homes have become inspirational symbols of modern living, offering valuable lessons and hope for future builders.

Lessons Learned and Advice Shared

Making a container home is more than just a construction project; it's a journey into sustainable and innovative living. Those who have embarked on this journey have accumulated a wealth of knowledge and insights that are invaluable to anyone considering a similar path. This comprehensive overview compiles experiences from numerous container home owners, providing a detailed account of the challenges they faced and the solutions they found.

Thorough planning emerges as a critical lesson from experienced container home builders. This stage sets the groundwork for a realistic budget, timeline, and project scope. Many owners stress the value of investing significant time in planning and consulting with architects, engineers, and other professionals who can offer guidance and foresee potential challenges. Detailed architectural plans are crucial not just for visualizing the space but also for navigating the complexities of building permits and zoning laws. Understanding and complying with local regulations is essential, as many homeowners have encountered costly setbacks when parts of their projects did not meet local standards.

The selection of shipping containers can significantly impact the success of the project. Homeowners suggest inspecting containers in person before purchasing to assess their condition and suitability for construction. While used containers are budget-friendly, they often come with issues like rust and structural wear that can increase refurbishment costs. Therefore, balancing cost with quality is crucial in selecting containers.

Container homes pose unique design and structural challenges. Maintaining the structural integrity of the home while making modifications for doors, windows, and other openings is vital. Many homeowners recommend enlisting experienced professionals, particularly for structural modifications, to ensure safety and compliance with building standards. Additionally, issues of insulation and ventilation are paramount in container homes due to their metallic nature and compact size. Effective insulation methods, such as spray foam, are frequently highlighted for their dual role in temperature control and moisture prevention.

Financial management is another area where container home owners have learned valuable lessons. Budget overruns are common, and planning for unexpected costs is crucial. Homeowners advise keeping a buffer of at least 10-20% over the estimated budget to accommodate unforeseen expenses. Engaging in DIY activities can save money, but homeowners caution against taking on tasks beyond one's skill level, which can lead to costlier fixes later.

The long-term maintenance of a container home requires attention to its unique aspects. Regular inspections for rust and corrosion are necessary to preserve the integrity of the steel structure. Homeowners have successfully used specialized paints and coatings to protect against the elements. Seasonal maintenance, particularly in extreme climates, is also crucial for preventing issues like moisture accumulation around the base of the home.

Finally, the support of a community that understands container home living can be incredibly beneficial. Many homeowners emphasize the benefits of connecting with others through forums, workshops, and container home events. These connections can offer support, new ideas, and a sense of community among like-minded individuals. The collective experiences of container home owners offer a roadmap for navigating the construction of these unique dwellings. From the planning phase through to long-term maintenance, the lessons learned are invaluable for new builders.

These insights not only help avoid common pitfalls but also enhance the overall success of container home projects, affirming the viability and sustainability of this innovative approach to housing.

The journey of building and living in a container home, as illustrated through the detailed case studies and success stories in this chapter, offers a profound insight into the transformative potential of shipping containers as sustainable housing solutions. These narratives not only chart the technical aspects of construction but also delve into the personal experiences, challenges, and triumphs of individuals who dared to think outside the traditional homebuilding box. From families seeking a return to simplicity and sustainability to individuals carving out urban retreats, each story adds to a collective wisdom that is invaluable for future builders.

The case studies presented demonstrate a wide array of approaches to designing, building, and living in container homes, each tailored to the specific needs and contexts of the owners. What stands out across all examples is the innovative use of space and materials that defines container home construction. Owners have pushed the boundaries of architectural design by transforming industrial containers into beautiful, functional homes that challenge conventional norms of residential construction. These homes are not merely structures; they are vibrant expressions of personal values and commitments to environmentally responsible living.

Each homeowner's journey underscored the necessity of resilience and creativity in overcoming the myriad challenges that come with unconventional projects like container homes. From navigating zoning laws and securing financing to tackling the physical demands of container modification and ensuring sustainable practices, the challenges are significant. However, as seen in the stories, these challenges are surmountable with thorough research, careful planning, and a willingness to learn and adapt. The proactive approach to learning from each obstacle and turning potential setbacks into opportunities for growth and improvement is a recurring theme that future builders can draw inspiration from.

The success stories highlighted the critical role of community and collaboration. Many homeowners emphasized the benefits of engaging with a network of like-minded individuals, from sharing advice and resources to offering moral support through the ups and downs of construction. Professional collaboration, too, proved vital, with architects, engineers, and contractors bringing specialized expertise that ensured the structural integrity, functionality, and aesthetic appeal of the homes.

These relationships underscore the importance of building a supportive ecosystem around the project, reinforcing the idea that creating a home is as much about building community as it is about construction.

Perhaps the most impactful aspect of these narratives is the wealth of practical knowledge and lessons shared by those who have lived through the process. Each story contributes to a growing repository of shared experiences that are gold mines for lessons on efficiency, sustainability, and design innovation. These lessons cover every phase of the project lifecycle, from the initial design and material selection to long-term maintenance and adaptation. For new builders, these lessons are not just technical guidelines but also philosophical insights into approaching a building project with an open mind and a flexible attitude.

The Future of Container Home Building

Looking forward, the stories in this chapter not only celebrate what has been achieved but also pave the way for future developments in container home building. As environmental concerns become increasingly urgent, the appeal of container homes as sustainable, efficient, and potentially cost-effective alternatives is likely to grow. Furthermore, as more people choose container homes and share their experiences, the collective knowledge and societal acceptance of such projects will expand, potentially influencing policy changes and more widespread adoption.

For readers contemplating their own container home projects, this chapter serves as both a manual and a source of inspiration. It illustrates that with enough determination, creativity, and respect for meticulous planning, turning the humble shipping container into a dream home is eminently achievable.

Each story is a testament to the spirit of innovation that drives the container home movement, providing not just practical blueprints but also encouraging a broader rethinking of what a home can be. In conclusion, the case studies and success stories of container homes presented in this chapter reflect a powerful blend of innovation, sustainability, and community. They provide not only a blueprint for building unique homes but also reflect a broader movement towards sustainable living practices worldwide.

As this book continues to explore the possibilities of container home building, it hopes to inspire, educate, and empower readers to embark on their own journeys of creating homes that embody a fusion of aesthetic beauty, practical function, and environmental stewardship. Each container home story is a building block in the evolving narrative of modern, sustainable architecture and a beacon for future explorations in the art of transforming spaces creatively and responsibly.

Made in the USA
Columbia, SC
02 December 2024